Television Symptom Diagnosis

Second Edition

An Entry Into TV Servicing

Television Symptom Diagnosis

Second Edition

An Entry Into TV Servicing

by

Richard W. Tinnell
Oklahoma State University

Howard W. Sams & Co.
A Division of Macmillan, Inc.
4300 West 62nd Street, Indianapolis, IN 46268 USA

SECOND EDITION
TENTH PRINTING — 1986

International Standard Book Number: 0-672-21460-1
Library of Congress Catalog Card Number: 77-83679

Printed in the United States of America.

Series Preface

Faced with a need for increasing numbers of technicians for an expanding electronics-radio-television industry, as well as with a rapidly expanding technology, the Consumer Electronics Group Service Committee of the Electronic Industries Association (EIA), in association with the Voorhees Technical Institute, Oklahoma State University, Appalachian State University, and various publishers, has long been active in developing and constantly revising educational materials to meet these challenges. In recent years, the pressing need for training programs to serve students of various backgrounds and abilities has also become evident.

These reasons, coupled with an increasingly critical shortage of trained technicians, especially in the home-entertainment electronics industry, have induced EIA to sponsor the preparation of a wide range of materials in three specific categories, tailored to meet a variety of requirements.

The following paragraphs explain these aids to education and suggest proper use of this material to achieve the desired results.

The laboratory text-manuals in the Basic Electricity Electronics Series provide in-depth, detailed, completely modern technical materials, combining a closely coordinated program of experiments, each preceded by a comprehensive discussion of the objectives, theory, and underlying principles. *Electricity-Electronics Fundamentals* provides an introductory course especially suitable for the preparation of service technicians or for other broad-based courses; *Basic Electricity* and *Basic Electronics* are planned as 270-hour courses, one to follow the other, providing a more thorough background for all levels of technician training. A related instructor's guide is available.

The Basic Electricity-Electronics Series

Title	Author	Publisher
Electricity-Electronics Fundamentals	Zbar/Sloot	McGraw-Hill Book Company
Basic Electricity	Zbar	McGraw-Hill Book Company
Basic Electronics	Zbar	McGraw-Hill Book Company

The Radio-Television Servicing Series includes materials in two categories: those designed to prepare apprentice technicians for performing in-home servicing and other apprenticeship functions, and those designed to prepare technicians for performing more sophisticated and complicated servicing such as bench-type servicing in the shop.

The apprenticeship servicing course includes *Television Symptom Diagnosis—An Entry into TV Servicing* (text, student workbook, instructor's guide) and *Television Symptom Diagnosis,*

The Radio-Television Servicing Series

Title	Author	Publisher
Television Symptom Diagnosis—An Entry into TV Servicing	Tinnell	Howard W. Sams & Co., Inc.
Television Symptom Diagnosis (33 film loops)	Tinnell	Howard W. Sams & Co., Inc.
Basic Radio: Theory and Servicing	Zbar	McGraw-Hill Book Company
Basic Television: Theory and Servicing	Zbar and Orne	McGraw-Hill Book Company

a series of 33 film loops. The first is a set consisting of profusely illustrated text, student response manual, and instructor's guide designed to train persons with no previous electronics training in job-entry troubleshooting skills for servicing in the home and shop. The text utilizes the "cue-response" concept of diagnosis, concentrates on identifying abnormal circuit operation and symptom analysis, and develops skills in troubleshooting. In the response manual, students are exposed to hundreds of television trouble symptoms through the use of color photos and illustrated problems. The instructor's guide is a complete and essential professional course of study that also contains the answers for the text and lab manual.

Television Symptom Diagnosis consists of a series of color-sound motion picture film loops, a student workbook, and an instructor's guide. These audiovisual materials function as an integrated learning system to teach color-television adjustment and setup procedures, trouble symptom diagnosis, and the ability to isolate troubles to a given stage in the receiver, concentrating on the requirements for servicing in the customer's home. But these audiovisual materials can also be used to supplement all levels of television courses. This medium is especially suitable for students who may have reading and, in turn, learning difficulties.

Since only a minimum of electronics theory is presented in the two courses described above, it is expected that apprentices completing these programs will be motivated to progress to the more comprehensive programs provided by the following courses in order to deepen their understanding of electronics and what makes the radio and television receiver work, and to obtain adequate proficiency for the servicing of all consumer electronics entertainment items.

The "bench-type" service technician courses consist of *Basic Radio: Theory and Servicing* and *Basic Television: Theory and Servicing*. These books provide a series of experiments, with preparatory theory, designed to provide the in-depth, detailed training necessary to produce skilled radio-television service technicians for both home and bench servicing of all types of radio and television. A related instructor's guide for these books is also available.

The Industrial Electronics Series

Title	Author	Publisher
Industrial Electronics	Zbar	McGraw-Hill Book Company
Electronic Instruments and Measurements	Zbar	McGraw-Hill Book Company

Basic laboratory courses in industrial control and computer circuits and laboratory standard measuring equipment are provided by the Industrial Electronics Series and their related instructor's guides. *Industrial Electronics* is concerned with the fundamental building blocks in industrial electronics technology, giving the student an understanding of the basic circuits and their applications. *Electronic Instruments and Measurements* fills the need for basic training in the complex field of industrial instrumentation. Prerequisites for both courses are *Basic Electricity* and *Basic Electronics*.

The foreword to the first edition of the EIA cosponsored basic series said: "The aim of this basic instructional series is to supply schools with a well-integrated, standardized training program, fashioned to produce a technician tailored to industry's needs." This statement is still the objective of the varied training program that has been developed through joint industry educator-publisher cooperation.

CHARLES N. HOFFMAN, SR. VICE-PRESIDENT
ELECTRONIC INDUSTRIES ASSOCIATION

Preface

In some cases, television receiver servicing demands a penetrating understanding of the nature of electromagnetic waves, composite video behavior, and the very nature of electronic theory. Fortunately, this is not so in the vast majority of problems.

Indeed, in almost all instances of set failure, a technician uses only four kinds of skill. These important areas of technical know-how are:

1. The ability to observe and recognize symptoms in the television picture and sound.
2. An understanding of the television receiver system which allows the technician to relate trouble symptoms to a particular section of the set.
3. A sufficient knowledge of electronic circuitry to allow isolation of the problem to an individual component.
4. The skill necessary to replace the defective part and restore normal operation.

The last one of these abilities involves skill in soldering and the use of basic hand tools which can only come from practice and experience.

The first three of these vital areas *can* be learned by almost anyone.

The fundamental purpose of this book is to present these three areas in a straightforward manner which will allow the reader to diagnose most television troubles. Toward this end, all of the material is of a nonmathematical and thoroughly practical nature.

The first two chapters can be considered an introduction to the manner in which a television set works. Each section of the receiver is discussed and its function explained.

Starting with Chapter 3, the individual stages of a typical set are considered. The essential principles of electronics are explained and applied to receiver servicing. In all cases the emphasis is on:

A. What constitutes NORMAL operation.

B. How abnormal operation can be identified and corrected.

As they become appropriate, the uses of electronic test instruments are presented. Simplicity and directness are the watchwords of these instrument discussions. Because of its usefulness and practicality, a special chapter on troubleshooting with the oscilloscope has been included before sync and sweep troubles. The oscilloscope (as well as other instruments) is relied upon in the chapter on color television troubleshooting.

Several chapters on color television are presented to provide an adequate coverage of color principles, setup, and color troubleshooting. Special features of modern receivers such as automatic fine tuning and remote controls are also included.

The final chapter is intended primarily as a quick reference of tv trouble diagnosis rather than as a review. This chapter can, however, be used as a summary.

It is worth mentioning that while considerable emphasis is on transistorized television circuitry, vacuum-tube circuits have by no means been ignored. Moreover, integrated circuits are also included where they are appropriate.

R. W. TINNELL

Acknowledgments

Acknowledgment is given to all persons and companies listed below for their assistance in the preparation and accomplishment of this EIA Vocational Technical Training Program.

EIA CONSUMER ELECTRONICS GROUP SERVICE COMMITTEE

C. John Borlaug
GTE Sylvania

George Camp, Jr.
Panasonic Company,
Division of Matsushita
Electric Corporation of America

Ed Gaiden
Quasar Electronics Company,
Division of Matsushita
Electric Corporation of America

Gerald N. Goldberger
Simpson Electric Company

Joe A. Groves
Howard W. Sams & Co., Inc.

Frank L. Hughes
Delco Radio,
Division of General Motors

Joe Hutton
Consumer Products Division
Arvin Industries, Inc.

John Kelly
Sony Corporation of America

Joseph V. Loiacono
General Electric Company

Brian J. Marohnic
Zenith Radio Corporation

Larry Menaugh
Rockwell International/Admiral

Walter H. Meyer
General Electric Company

Dick Pavek
Tech Spray

Stephen Phillips
Harman International

Jim Scallions
RCA Consumer Electronics

Ed Schlenner
GTE Sylvania Service Co., Inc.

Norman F. Schumacher
Wells-Gardner Electronics Corp.

R. J. Sokolowski
RCA Service Company

Ray Yeranko
Magnavox Consumer Electronics
Company

Subcommittee on Manual Revisions

Joe A. Groves
Howard W. Sams & Co., Inc.

Ed Milbourne
RCA Consumer Electronics

Sid Baker
RCA Service Company

Contents

CHAPTER 6

CHAPTER 7

CHAPTER 8

CHAPTER 9

CHAPTER 10

CHAPTER 11

CHAPTER 12

CHAPTER 13

CHAPTER 14

CHAPTER 15

CHAPTER 16

CHAPTER 17

CHAPTER 18

CHAPTER 1

Introduction to Television

People have always wanted to see things for themselves, and almost everybody agrees that it is a lot better to see a thing than just to hear or read about it. That is probably why television has been so widely accepted by the public.

Stop and think about it. It really is miraculous that you can see and hear things happening anywhere in the world and even in outer space.

To produce this miracle, a television receiver actually does only two things: it shows a picture, and it delivers the accompanying sounds to be heard. Consider for a moment how a tv set does these things.

1-1 Scanning the Television Picture

Suppose you were talking on the telephone and you wanted the person at the other end to get a good idea of what this printed page looked like. You would probably start the description at the upper left corner and go along the first line from left to right. At the end of the line you would jump back to the left and down to the second line. In this way you would proceed through the page until you got to the lower right corner. If you wanted to repeat the description you would jump back to the upper left and go through the whole process again.

A television station does just about this same thing for each picture it sends out. For example, look at the simple picture in Fig. 1-1. The tv camera "looks" at each little square one at a time and transmits what it finds there to the receiver. In doing so, the camera scans from left to right and from top to bottom just like the foregoing description of the printed page. Each of the squares is called a *picture element.* An actual tv picture contains about 150,000 of these elements rather than only about 200 as shown in the illustration.

It is obvious that the more picture elements there are the better the picture quality will be. For example, the eye of the dog in Fig. 1-1 would probably not show up in the tv picture because it is very small compared to the size of a square, and it is spread among several different squares. On the other hand, if there were a thousand times as many squares (picture elements), the eye, all by itself, would occupy many squares. This ability of a picture to show small details clearly is called its *resolution.*

In order for the picture to be reproduced properly, the television receiver must scan its screen exactly in step with the camera's scanning of the scene. To ensure that the receiver and camera are in step, a set of impulses is supplied to the camera to tell it when to start at the top of the scene and when to start each line at the left side of the scene. These same impulses are sent to the receiver along with the picture-element information. These impulses are called *synchronizing (sync) signals.* The ones that tell the camera and receiver to start at the top are termed *vertical-sync pulses,* and those that start the scanning of each line at the left are known as *horizontal-sync pulses.*

A television station in the United States sends out 30 complete pictures every second. Each one of these pictures is called a *frame* and consists of 525 horizontal *lines.* Each line contains about 285 picture elements, which explains the excellent picture quality seen on a television set.

With only 30 pictures per second, there is a noticeable flicker like that seen in an old-time silent movie. To reduce this flickering, the television system actually scans 262½ lines spread evenly down the height of the screen (shown in

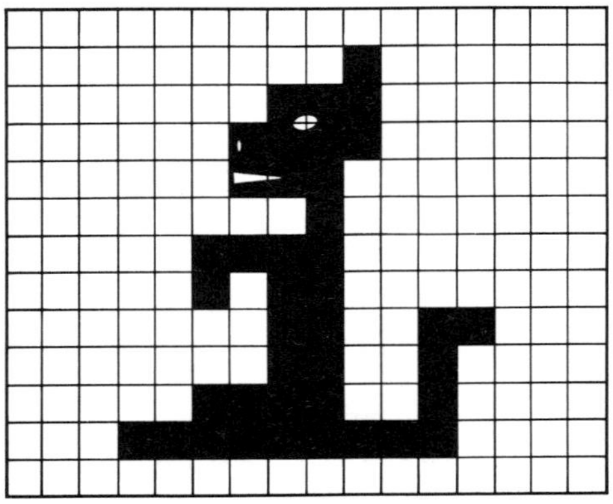

Fig. 1-1. Basic structure of a tv picture.

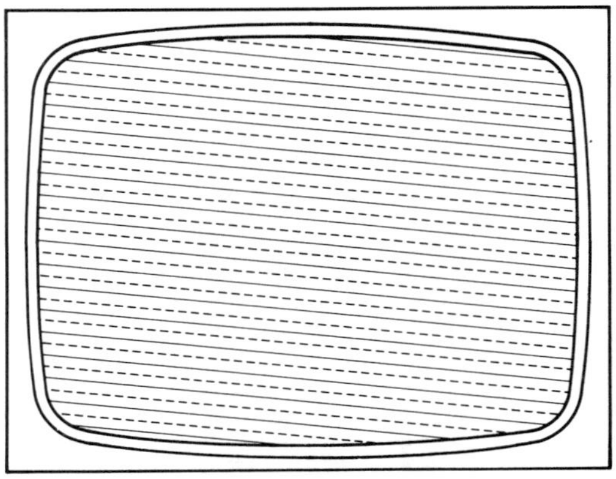

Fig. 1-2. Interlaced scanning.

solid lines in Fig. 1-2) then goes back and scans another 262½ lines between the first ones. (The second scan lines are dashed in Fig. 1-2). This method of reducing flicker is called *interlaced scanning*. It gives much the same effect as having 60 complete pictures per second. Each set of 262½ lines is called a picture *field*. It therefore takes two fields to produce one complete frame.

If a television receiver is operating and there is no picture information present, then the set simply reproduces the scanning pattern. This pattern consists of horizontal white lines and is called the scanning *raster*. The presence or absence of the raster is a very useful piece of troubleshooting information. A more detailed consideration of the raster as a troubleshooting aid will be presented later. In the meantime, the raster produced by a set can be examined by disconnecting the antenna wire and switching to an unused channel. A pattern of white horizontal lines should appear.

1-2 Picture Information

When a television camera looks at a picture element, it "sees" several pieces of information. By the use of a special optical system, the camera determines the amount of *red, green,* and *blue* color in the element as well as its *brightness*. This information is sent to the receiver, where the picture element is reproduced by combining correct proportions of red, green, and blue. (In a black-and-white set, only the brightness of the element is reproduced.) The brightness signal and the red, green, and blue signals that are sent to the receiver are called the *video* signals. The signal that coordinates the color information is called the *color burst*.

Both the black-and-white signal and the color signals may be observed at the receiver, and this information is very helpful in determining the cause of a failure. How a service technician uses observations of these signals to diagnose problems will be described later.

1-3 Television Sound

The sound that accompanies a tv picture is also very important. In general, the sound (*audio*) signal is very much like an ordinary fm radio signal. In Chapters 2 and 3, how the sound system works and how to use the sound to help locate troubles in a receiver will be examined in detail.

At this point, observe that one vital factor not present in fm radio must be satisfied. That is, the sound must be "tuned in" at the same time that the picture is.

1-4 Television Channels

Each authorized television station produces the three kinds of information described above (picture, sound, and sync) and transmits them on an assigned channel. This is done by superimposing the information on a *picture carrier* signal and a *sound carrier* signal. These carrier signals alternate from electrically positive to electrically negative many millions of times a second. The exact number of times per second a signal alternates is called the *frequency* of the signal. The basic unit of frequency is the *hertz*. For example, a signal that alternates 3000 times per second has a frequency of 3000 hertz. Because television carriers have frequencies in the millions of hertz, these carriers are usually given in terms of *megahertz* (one megahertz is one million hertz). For example, a frequency of 54,000,000 hertz would usually be given as 54 megahertz. The term hertz is abbreviated Hz, and megahertz is abbreviated MHz.

Table 1-1 lists the frequencies for the vhf tv channels. The uhf channels start with channel 14, which begins at 470 MHz, and extend through channel 83, which ends at 890 MHz.

Table 1-1. VHF Television Channels

Channel Number	Frequency Limits (MHz)	Picture Carrier (MHz)	Sound Carrier (MHz)
2	54-60	55.25	59.75
3	60-66	61.25	65.75
4	66-72	67.25	71.75
5	76-82	77.25	81.75
6	82-88	83.25	87.75
7	174-180	175.25	179.75
8	180-186	181.25	185.75
9	186-192	187.25	191.75
10	192-198	193.25	197.75
11	198-204	199.25	203.75
12	204-210	205.25	209.75
13	210-216	211.25	215.75

1-5 Summary

For normal operation, a modern television receiver must receive three types of signal. They are:

1. The synchronizing pulses
2. The picture information
3. The accompanying sound

With these signals, the set must produce:

1. A scanning raster that is synchronized with the camera
2. A picture that has correct brightness and color
3. Sound that is coordinated with the picture

1-6 Review Questions

1. What is meant by each of the following terms:

   A. Scanning?
   B. Picture element?
   C. Resolution?
   D. Sync?
   E. Frame?
   F. Field?
   G. Interlaced scanning?
   H. Audio?
   I. Line?
   J. Raster?
   K. Video?
   L. Color burst?

2. How many horizontal lines make one frame? One field?
3. About how many picture elements are there in a frame? A field? A line?
4. What is the principal advantage of using interlaced scanning in the television system?
5. What is the relationship between picture resolution and picture detail?
6. What are the two types of sync pulses required by a tv set?
7. What are the two types of picture information supplied to a tv set?
8. How many fields are produced per second by a tv set? How many frames?
9. How many times per second does a tv set scan from top to bottom? How many times from left to right?
10. What are the three basic colors used in color tv?

CHAPTER 2

The Television Set—Block Diagrams

Block diagrams are extremely useful to the television technician. They help him organize his thinking about what goes on in a set, and they help him locate troubles within the set.

Before examining the block diagram of a typical set, think about this whole idea of representing a tv in block form. A single block could be used to represent a whole tv set. Fig. 2-1 shows such a block diagram. While this is a simple and easy way to represent the set, it does not tell much about the operation of the set.

To make the block diagram more useful, some of the things that the set does can be included in the diagram. First the set must have a source of electric power—usually 120 volts from an ac wall outlet. Then too, the set must have some kind of antenna to pick up the signal from the station. These two things (power and signal) are the *inputs* required by the set. On the *output* side are the picture and the accompanying sound. If these inputs and outputs are added pictorially to the block diagram, one like Fig. 2-2 will result.

While this block diagram is more informative than the first one, it still does not tell much about what goes on inside the set. What must be done at this point is to put in blocks that represent the basic functions of the set.

In Chapter 1, it was shown that the things that a tv receiver must do are:

1. Produce a scanning raster that is synchronized with the station signal.
2. Paint a picture that is of correct brightness and color.
3. Reproduce the sound that is associated with the picture.

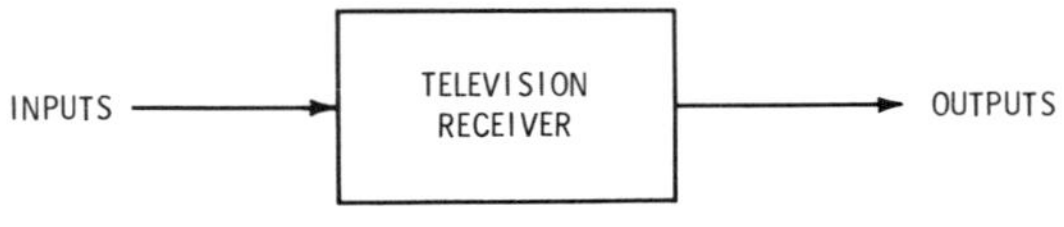

Fig. 2-1. An elementary tv block diagram.

To do these things, a normal tv set also provides three other functions. It selects the desired channel from among the several available ones. Then, since the station signal is usually very weak, the set *amplifies* the signal up to a usable level. Finally, the set converts the household electricity into a form usable within the set.

If all these things are included in the block diagram, it will look something like Fig. 2-3. This diagram, while still relatively simple, does begin to show how the set works. Notice that the flow of signal can be followed through the set. For instance, the signal from the station first enters the antenna. Then the *tuner* selects the desired channel. Next, the *if* (intermediate-frequency) *amplifiers* amplify only the chosen signal. After amplification, the signal is split into three parts (sound, picture, and raster), which are processed separately, but all at the same time. While all this is

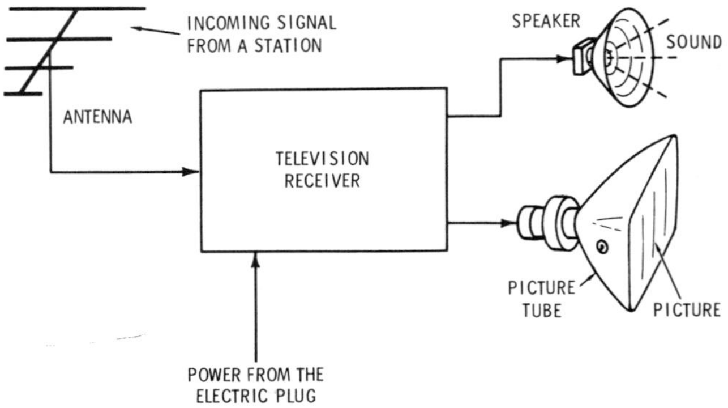

Fig. 2-2. Block diagram showing inputs and outputs.

SAMS™

Book Mark

Sams books cover a wide range of technical topics. We are always interested in hearing from our readers regarding their informational needs. Please complete this questionnaire and return it to us with your suggestions. We appreciate your comments.

1. Which brand and model of computer do you use?

☐ Apple ____________
☐ Commodore ____________
☐ IBM ____________
☐ Other (please specify) ____________
____

2. Where do you use your computer?

☐ Home ☐ Work

3. Are you planning to buy a new computer?

☐ Yes ☐ No

If yes, what brand are you planning to buy? ____________

4. Please specify the brand/type of software, operating systems or languages you use.

☐ Word Processing ____________
☐ Spreadsheets ____________
☐ Data Base Management ____________
☐ Integrated Software ____________
☐ Operating Systems ____________
☐ Computer Languages ____________

5. Are you interested in any of the following electronics or technical topics?

☐ Amateur radio
☐ Antennas and propagation
☐ Artificial intelligence/expert systems
☐ Audio
☐ Data communications/telecommunications
☐ Electronic projects
☐ Instrumentation and measurements
☐ Lasers
☐ Power engineering
☐ Robotics
☐ Satellite receivers

6. Are you interested in servicing and repair of any of the following (please specify)?

☐ VCRs ____________
☐ Compact disc players ____________
☐ Microwave ovens ____________
☐ Television ____________
☐ Computers ____________
☐ Automotive electronics ____________
☐ Mobile telephones ____________
☐ Other ____________

7. How many computer or electronics books did you buy in the last year?

☐ One or two ☐ Three or four
☐ Five or six ☐ More than six

8. What is the average price you paid per book?

☐ Less than $10 ☐ $10-$15
☐ $16-$20 ☐ $21-$25 ☐ $26+

9. What is your occupation?

☐ Manager
☐ Engineer
☐ Technician
☐ Programmer/analyst
☐ Student
☐ Other ____________

10. Please specify your educational level.

☐ High school
☐ Technical school
☐ College graduate
☐ Postgraduate

11. Are there specific books you would like to see us publish? ____________
____
____

Comments ____________
____
____

Name ____________
Address ____________
City ____________
State/Zip ____________

21460

SAMS™

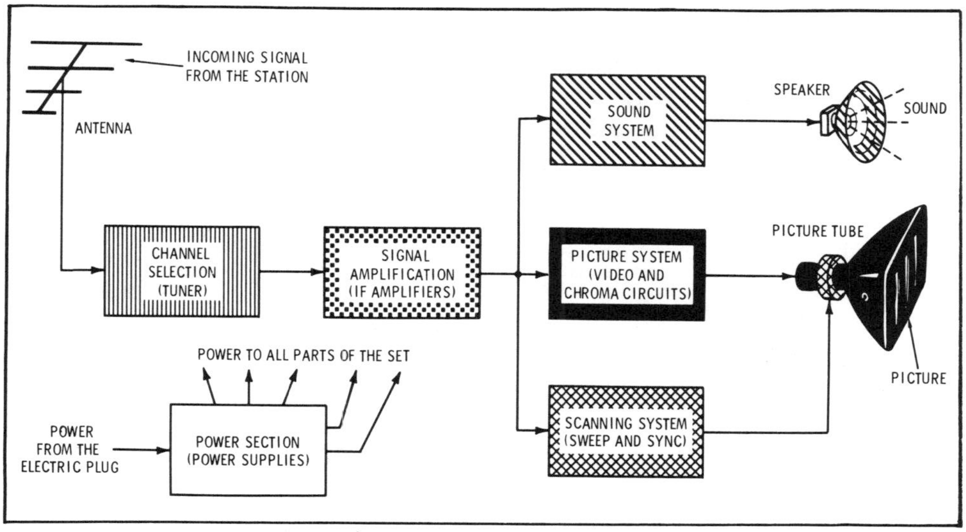

Fig. 2-3. A basic functional block diagram.

going on, the *power supplies* are providing the required power to all sections of the set. This type of block diagram is called a *functional block diagram* because it shows what the set does and the sequence in which it does it.

Even though the diagram in Fig. 2-3 is functional, it does not contain enough detail to be really useful in troubleshooting a set failure. Therefore, examine the typical block diagram shown in Fig. 2-4. The blocks have been shaded

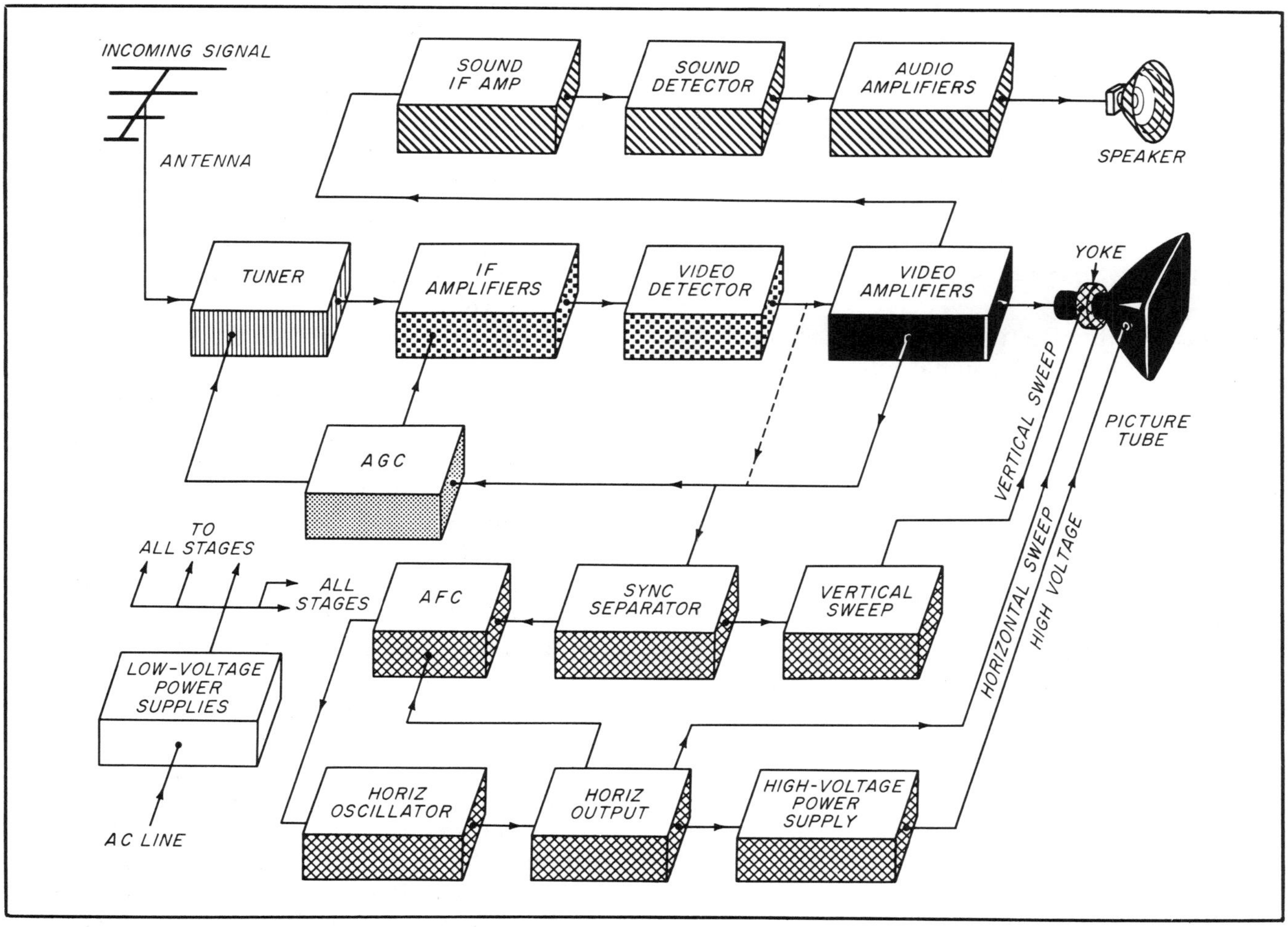

Fig. 2-4. Block diagram of a typical black-and-white receiver.

to help in relating this complex diagram to the simpler one in Fig. 2-3. Also, the color (chroma) blocks have been left out. The color section will be considered in some detail in Chapter 14.

Now, examine the diagram (Fig. 2-4) block by block and see just what each one does.

2-1 The Antenna

Technically speaking, the antenna (Fig. 2-5) is part of the television tuner. However, since it is separate from the set, it is often treated as an independent block. The basic function of the antenna is to pick up the signal from the station and apply it to the tuner input.

In a practical case, the antenna should be able to work with any desired channel. You may discover antennas that work only on the *vhf channels* (channel 2 through channel 13), or you may find some that are designed for only the *uhf channels* (channel 14 through channel 83). In addition, there are many that are used for all of the channels, as well as some that are designed for only one channel.

2-2 The Tuner Block

Since there are 83 television channels altogether and it is desired to see only one of them at a time, there must be a way to keep the other 82 channels out of the set. This job is done by the tuner (Fig. 2-6).

In addition to selecting the desired channel, the tuner also converts the signal into a form usable by the if amplifiers. To understand the need for this conversion, examine the tuner signal in more detail.

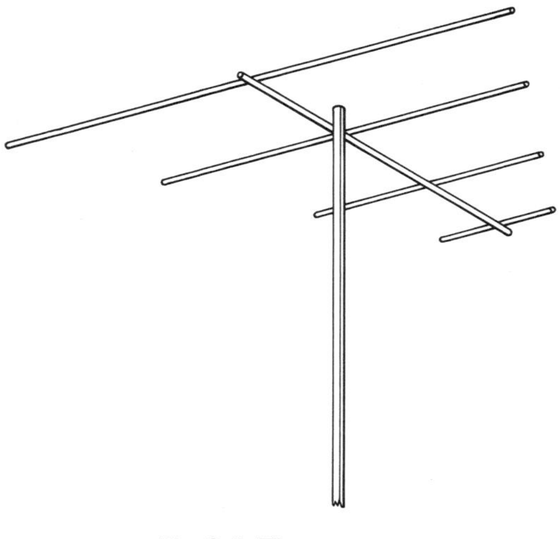

Fig. 2-5. The antenna.

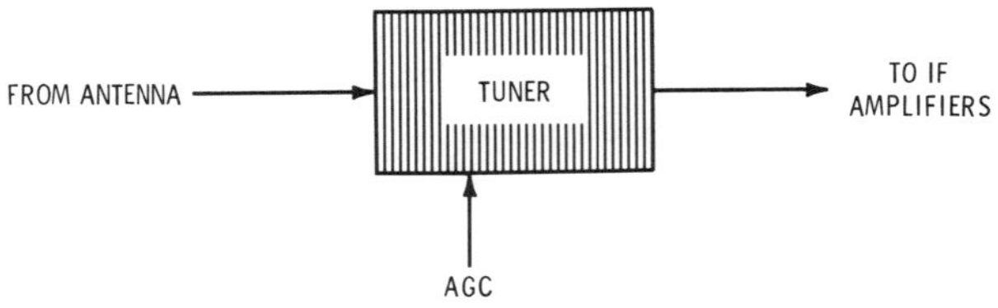

Fig. 2-6. The tuner.

The tuner receives signals from one or more stations. Each station transmits on its own channel. It has its own power level and may be a different distance from the set.

If it were necessary to amplify all channel frequencies with the if amplifiers, each amplifier would have to be retuned every time the channel was changed. This would be very involved and expensive. Instead, what is done is to adjust the if amplifiers to one very special channel. This special channel is called the *if band*. Then, in the tuner, whatever channel is selected is converted to the if band of frequencies. It may be helpful to think of the if band as channel number one. While this is not strictly true, it is near enough to being true to be a useful idea.

Now that the signal is ready for the if amplifier, its strength must still be considered. It would be ideal for the set to produce the same high-quality picture and sound for a weak (distant) station as it does for a strong (near) station. For this to happen, the signal strength must be adjusted in the tuner. This adjustment is made automatically by a control signal that is called the *agc* (*automatic gain control*). The agc will be examined in more detail in a later section.

In summary, the tuner does three things:

1. It selects the desired channel.
2. It converts the signal to the if band.
3. It automatically adjusts the strength of the desired signal.

As you might guess, the *channel selector* and *fine-tuning control* are in the tuner section.

2-3 The IF Amplifiers

Under normal conditions, the signal from the tuner section of a television receiver is not nearly strong enough to operate the picture and sound sections. Therefore, it is necessary to increase the signal strength. The basic purpose of the if (intermediate-frequency) amplifiers (Fig. 2-7) is to provide the required signal gain.

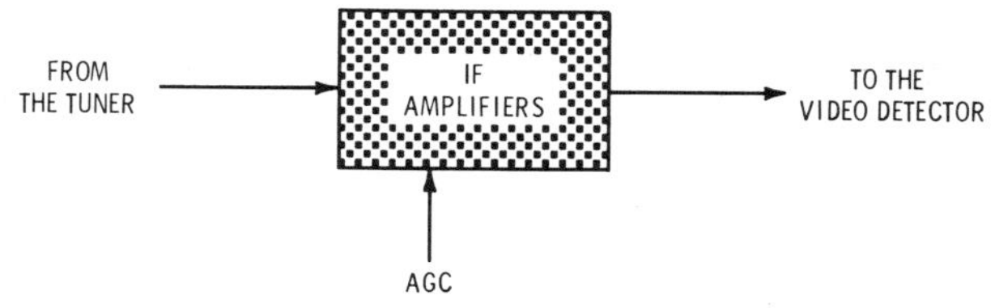

Fig. 2-7. The if amplifiers.

Automatic gain control (agc) is normally applied to at least one of the if amplifiers to help control the signal level from the tuner so that the set can produce equally good results with both strong and weak station signals.

2-4 The Video Detector

From the if amplifiers the signal is sent to the video detector (Fig. 2-8). The signal is now large

Fig. 2-8. The video detector.

enough to operate the rest of the set. So, the video detector converts the signal from the if amplifier and produces two kinds of output. The outputs from the detector, which are both fed to the video amplifiers, are:

1. The *composite video* signal, which contains the picture information and the synchronizing pulses
2. The *sound if* signal, which is an fm radio signal carrying the sound information

2-5 The Video Amplifiers

In the video-amplifier stages (Fig. 2-9), the video-detector output is amplified first. Then the sound if signal and the synchronizing pulses are separated from the picture information and are passed on to their individual sections.

The sound if signal is directed to the sound section of the set. The synchronizing pulses are supplied to both the agc and the sync separator. The actual picture information (video) is further amplified and sent to the picture tube. The *contrast control* is located in this section. It controls the amount of video sent to the picture tube, much like the volume control adjusts the audio level to the speaker.

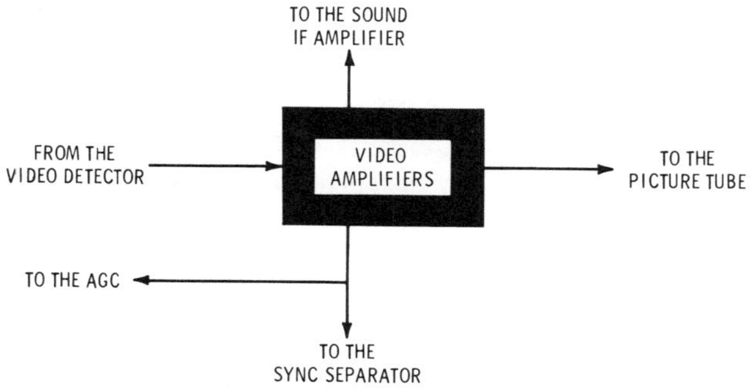

Fig. 2-9. The video amplifiers.

In the case of a color set, the color signals are separated at this point and processed by the color section of the receiver. The color section will be covered in Chapter 14.

2-6 Automatic Gain Control

The agc system supplies a control voltage to the tuner and if amplifier blocks. The purpose of this voltage is to produce a good picture for both strong and weak station signals.

The function of the agc block (Fig. 2-10) is to monitor the average level of the synchronizing pulses. If their level is low, the agc automatically increases the amplification of the tuner and the if amplifiers. If the level of the synchronizing pulses is too high, then the agc decreases the amplification of the tuner and the if amplifiers. In this way the system tends to hold the signal at a level that will produce good-quality pictures.

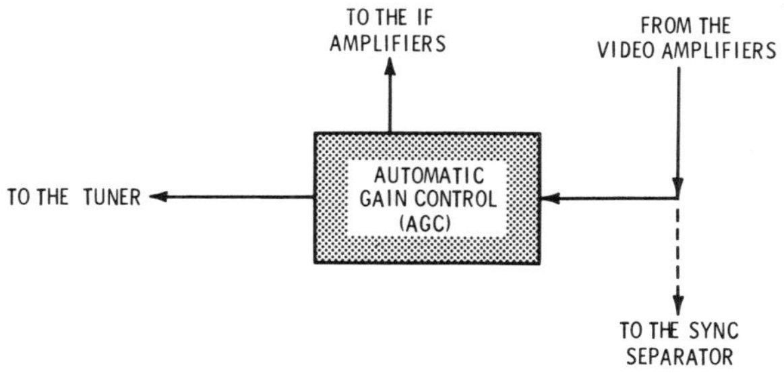

Fig. 2-10. The automatic gain control.

2-7 The Sound IF Amplifier

The sound if amplifier (Fig. 2-11) has a simple function. It amplifies the fm radio signal that carries the sound information.

2-8 The Sound Detector

The fm radio signal from the sound if amplifier is not suitable for reproduction in the speaker, so

Fig. 2-11. The sound if amplifier.

it is necessary to separate the sound information from this carrier. This is done by the sound detector (Fig. 2-12A). The output of the detector is an electrical audio signal much like that from a phonograph pickup.

(A) The sound detector.

(B) The audio amplifiers.

Fig. 2-12. The sound system.

2-9 The Audio Amplifier

In the case of a phonograph, the audio signal from the pickup must be amplified before it can operate the speaker. Exactly the same thing is true in a tv set. The audio from the sound detector must be amplified before it can operate the speaker. In a tv receiver, the section that does this amplification is called the *audio amplifier section* (Fig. 2-12B). As in a phonograph, the *volume control* will be found in this section. In many sets, there is also a *tone control* in the audio section.

2-10 The Speaker

The speaker is normally considered a part of the audio system. Its function, of course, is to convert the electrical audio impulses into audible sound.

2-11 The Sync Separator

The synchronizing pulses from the video amplifiers are applied to the sync separator as well as to the agc. The sync separator (Fig. 2-13) separates the sync pulses from the video and sends

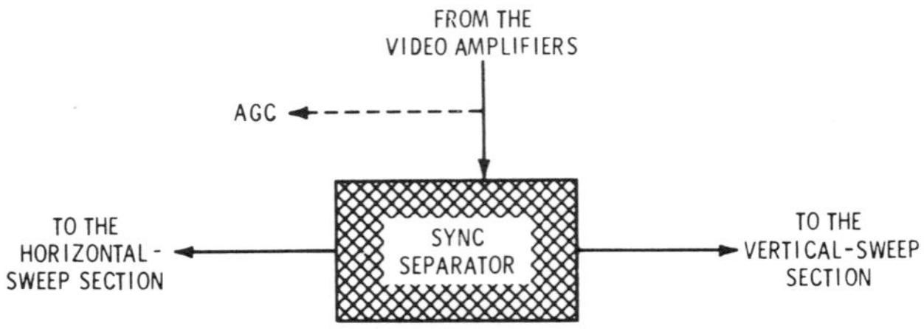

Fig. 2-13. The sync separator.

only the vertical ones to the vertical-sweep section and only the horizontal ones to the horizontal sweep circuits.

2-12 The Vertical-Sweep Section

The vertical-synchronizing pulses are applied to the vertical-sweep section (Fig. 2-14). In this portion of the set, the vertical-sweep signal is produced and applied to the deflection yoke. This signal causes the beam of the picture tube to sweep vertically at a rate set by the sync pulses. As a result, the picture will be synchronized vertically with the scene being telecast by the station.

Fig. 2-14. The vertical-sweep section.

In this section, the *vertical-hold control* as well as controls for varying the picture *height* and *linearity* will be found.

2-13 The Horizontal-Deflection Oscillator

The horizontal-synchronizing pulses from the sync separator are applied first to the afc (automatic frequency control) and then to the horizontal oscillator (Fig. 2-15). Further coverage of the afc will be given in Section 2-15. In the meantime, assume that the horizontal-sync pulses reach the oscillator block.

Fig. 2-15. The horizontal oscillator.

The horizontal oscillator produces a horizontal-sweep signal in step with the sync pulses supplied to it. The sweep signal is then directed to the horizontal-output stage. In the horizontal-oscillator section, the *horizontal-hold control* will be found.

2-14 The Horizontal-Output Section

It takes 525 horizontal lines to produce each individual picture. Therefore, the picture-tube spot must move back and forth very rapidly. Actually, it moves across the screen and back 15,750 times every second. To move the spot so fast requires a considerable amount of horizontal-sweep power. This power is provided by the horizontal-output section (Fig. 2-16) of the set.

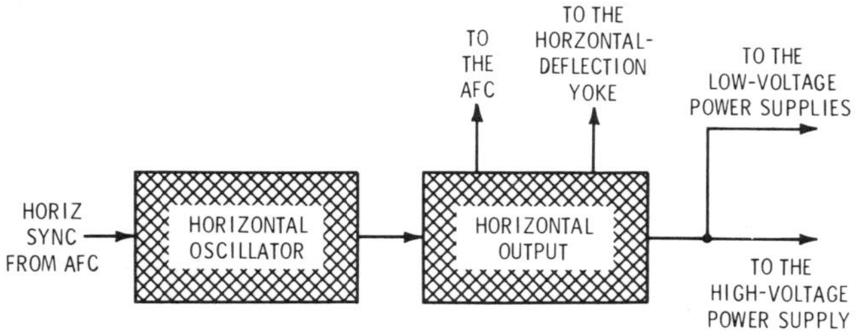

Fig. 2-16. The horizontal-output section.

The horizontal-output section is a powerful pulse generator which is triggered by the horizontal oscillator. The main job of the horizontal-output block is to provide adequate horizontal-sweep power to the horizontal-deflection yoke. Within the horizontal-output section will normally be found a picture *width control* and sometimes a *horizontal-linearity control.*

In addition to providing power to sweep the spot horizontally, the output section also powers the high-voltage power supply. The output block also sends a pulse back to the afc section. This will be considered next.

2-15 Horizontal AFC

As mentioned before, the horizontal-output section is a powerful pulse generator which is triggered by the horizontal oscillator. The rate at which the horizontal-output stage operates determines the actual rate at which the picture is scanned, and it must be kept in step with the frequency at which the transmitted signal is sent.

To ensure that the horizontal output is in step with the sync pulses from the station, the two are compared. Both the station sync pulses and the horizontal-output pulses are fed to the afc (automatic frequency control) section (Fig. 2-17). The afc circuit in turn controls the horizontal oscillator. If the horizontal-output pulse rate is too fast, the afc slows down the horizontal oscillator. This in turn slows down the horizontal-output rate.

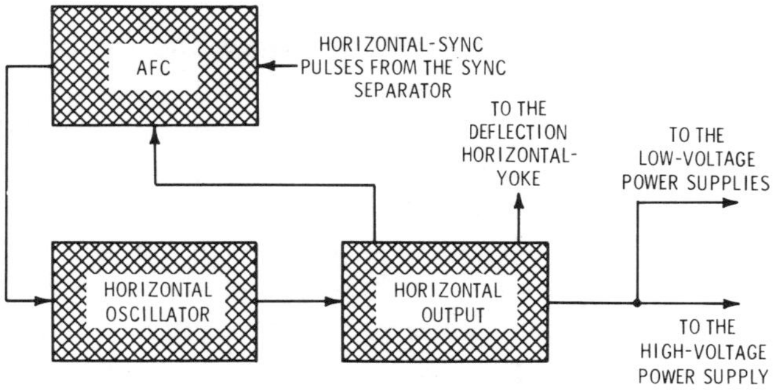

Fig. 2-17. The horizontal afc system.

If the horizontal-output pulse rate is too slow, the afc speeds up the horizontal oscillator, which causes the horizontal-output rate to increase. In this way, the afc section holds the horizontal sweep rate at the correct value.

2-16 The High-Voltage Power Supply

In addition to providing for the sweep and afc, the horizontal-output section also feeds the high-voltage power supply (Fig. 2-18). This power supply takes the high-powered pulses from the output stage and converts them into a high voltage (usually over 12,000 volts) to operate the picture tube.

Fig. 2-18. The high-voltage power supply.

2-17 The Low-Voltage Power Supplies

Every part of the television set requires power to operate properly. The function of the low-voltage power supply (Fig. 2-19) is to take the power from the electric wall outlet and convert it to the form needed by the various set sections. The power supply normally includes the *on-off switch* that controls all of the power to the set.

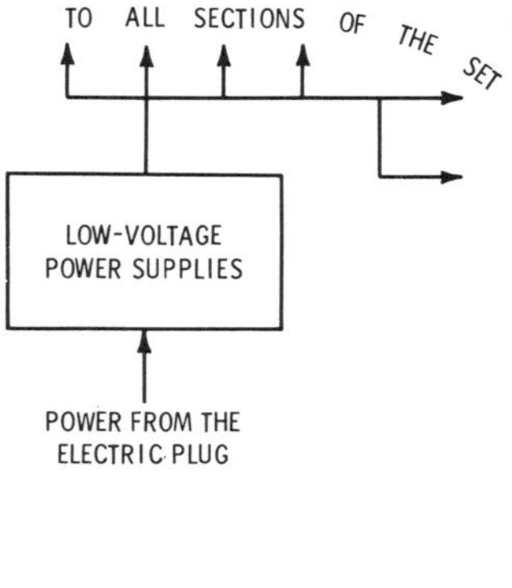

Fig. 2-19. The low-voltage power supplies.

This section also normally includes fuses and circuit breakers to protect both the set and the home from damage due to excessive current.

Many recent sets include one or more additional low-voltage supplies that get their primary energy from some other section, usually the horizontal-output circuit. These low-voltage supplies work in the same way as power-line–operated ones, but in general, they use smaller, more compact parts. The use of this type of power supply reduces the energy requirements of the set.

2-18 The Picture Tube

The picture tube (Fig. 2-20) receives four different inputs from the set. They are:

1. High voltage to operate the picture tube
2. Vertical sweep, which controls the rate that the picture is scanned from top to bottom
3. Horizontal sweep, which causes the picture to be scanned from side to side

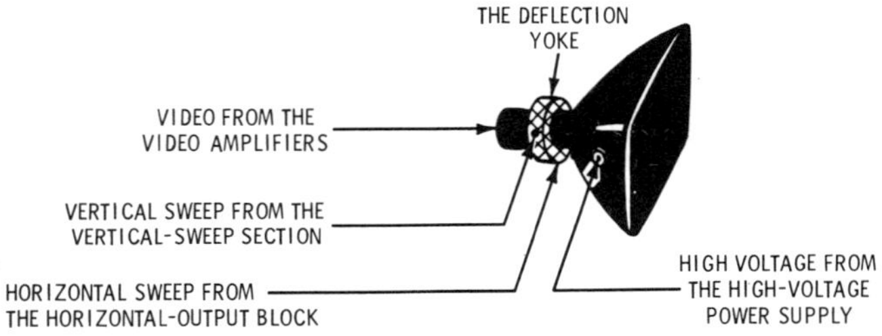

Fig. 2-20. The picture tube.

4. Video, which varies the intensity of the picture as it is scanned

With these four inputs, the picture tube reproduces the scene being telecast by the station. Also associated with the picture tube are the *focus* and *brightness* controls.

2-19 Modular Construction

Modular type construction is used in a great many current sets. In this type of construction, circuit functions are contained on a group of circuit boards called *modules.* (In some receivers, different functions can appear on a single module.)

Fig. 2-21 shows a *module diagram* of a typical modern set. Notice how similar the module diagram is to the functional block diagram given in Fig. 2-3. In many cases, module diagrams can be used very effectively in troubleshooting problems.

2-20 Summary

Television receiver block diagrams can be very useful to the service technician. They allow him to trace the path that the station signal takes through the set, and they can be helpful if the set fails to work properly. The technician can usually tell which part of the set has stopped working just by observing the external symptoms.

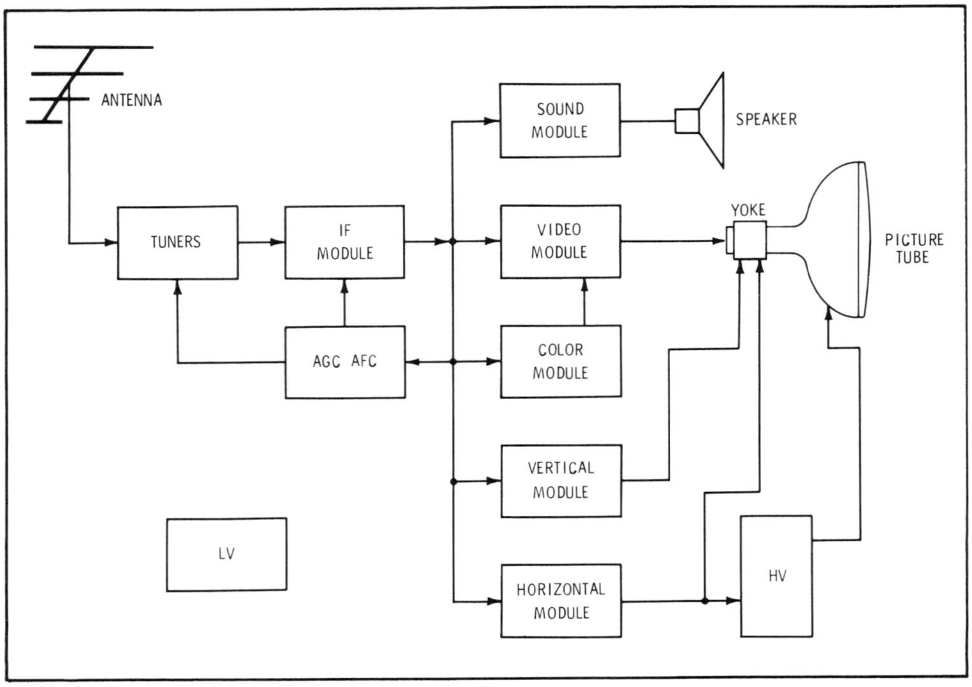

Fig. 2-21. Diagram of a modular tv receiver.

Fig. 2-4 is a typical television receiver block diagram. It is perhaps worth mentioning at this point that block diagrams are not normally shaded. However, the use of different shading here will be helpful in identifying the various major functions of the set.

In the illustration, the tv signal enters the set at the antenna and goes to the tuner. Within the tuner, the desired channel is selected, the signal is converted to the if band, and the level is adjusted. The if signal is amplified by the if amplifiers. The signal is converted to an fm radio signal (sound) and a composite video signal (picture and sync) by the video detector. These signals are amplified and sent to the correct sections by the video amplifiers. The video is applied to the picture tube. The fm sound signal is amplified, converted to audio, amplified some more, and then sent to the speaker. The sync signal goes to the sync separator.

The sync pulses are used to control the level of the signal at the tuner and if amplifier by means of the agc. Also, the sync pulses are separated, vertical from horizontal, and sent to the sweep circuits. The vertical-sweep system controls the picture-tube scan in the vertical direction. The horizontal sweep is produced by the afc, horizontal oscillator, and horizontal-output sections. The output section is synchronized with the station sync pulses through the action of the afc and the horizontal oscillator. The horizontal-output stage also feeds the picture-tube high-voltage power supply.

The low-voltage power supplies furnish the required operating power to all sections of the set.

Before proceeding, you should study the block diagram shown in Fig. 2-4 until you have the overall set operation firmly in mind.

2-21 Review Questions

1. Explain the path that the sound signal takes from the antenna to the speaker.
2. In which blocks are the sound and picture information both present?
3. What are the functions of the tuner?
4. Explain how the horizontal sweep at the picture tube is synchronized with the sync pulses from the station.
5. What is the function of the sync-separator block?
6. Which block in Fig. 2-4 processes *only* the sound signal?
7. Suppose you know that there are three amplifier stages in the if amplifier block and that agc is applied only to the one nearest the tuner. Draw a block diagram showing such an if system.
8. Which blocks in Fig. 2-4 process both the vertical- and horizontal-sync pulses?
9. Explain how the signal level is adjusted as the set is changed from a strong station to a weak one.
10. Which block has an effect on every part of the set?
11. Suppose that you know that the vertical-sweep section contains two stages: a vertical oscillator that is synchronized with the station, and a vertical output. Draw a block diagram of the section.
12. Which blocks in Fig. 2-4 process *only* the horizontal signals? Which only the vertical signals?
13. Which parts of the set must be able to handle signals from all 82 channels?
14. What is the purpose of the agc system?
15. If there are three amplifier stages within the video-amplifiers block, and the sync, agc, and sound are taken out after the first video stage, what would the block diagram of the video stages look like? (Draw the block diagram.)
16. What is the function of the video detector?
17. Which block in Fig. 2-4 processes only the picture-tube high voltage?
18. What does the horizontal-output stage do?
19. List the blocks through which the horizontal-sync pulses *must* pass.
20. In which block is the contrast control located?
21. In which block would you expect to find each of the following controls?

    A. Volume
    B. Focus
    C. Brightness
    D. Width
    E. Contrast
    F. On-off
    G. Vertical hold
    H. Tone
    I. Fine tune
    J. Height
    K. Horizontal hold
    L. Vertical linearity
    M. Channel selector

22. List the blocks through which the picture

information must pass to get from antenna to picture tube.

23. Explain the functions of the video amplifiers.
24. Which of the following takes place first:

    A. Amplification of the composite video?

    B. Separation of the fm sound signal from the composite video?

    C. Separation of the horizontal- and vertical-sync pulses from each other?

25. Explain the function of the horizontal-oscillator/output section.

CHAPTER **3**

Audio-Section Failures

When audio signals in a television receiver are discussed, two different things are involved. The *sound* that comes from the speaker is called an audio signal. And, the *electronic impulses* that produce the sound are also called an audio signal. In this chapter, the operation of the electronic audio section of a television receiver will be examined. Moreover, several types of audio failures will be considered.

3-1 Identifying Audio Failure Symptoms

The whole set block diagram shown in Fig. 3-1 has the audio sections shaded. These sections are

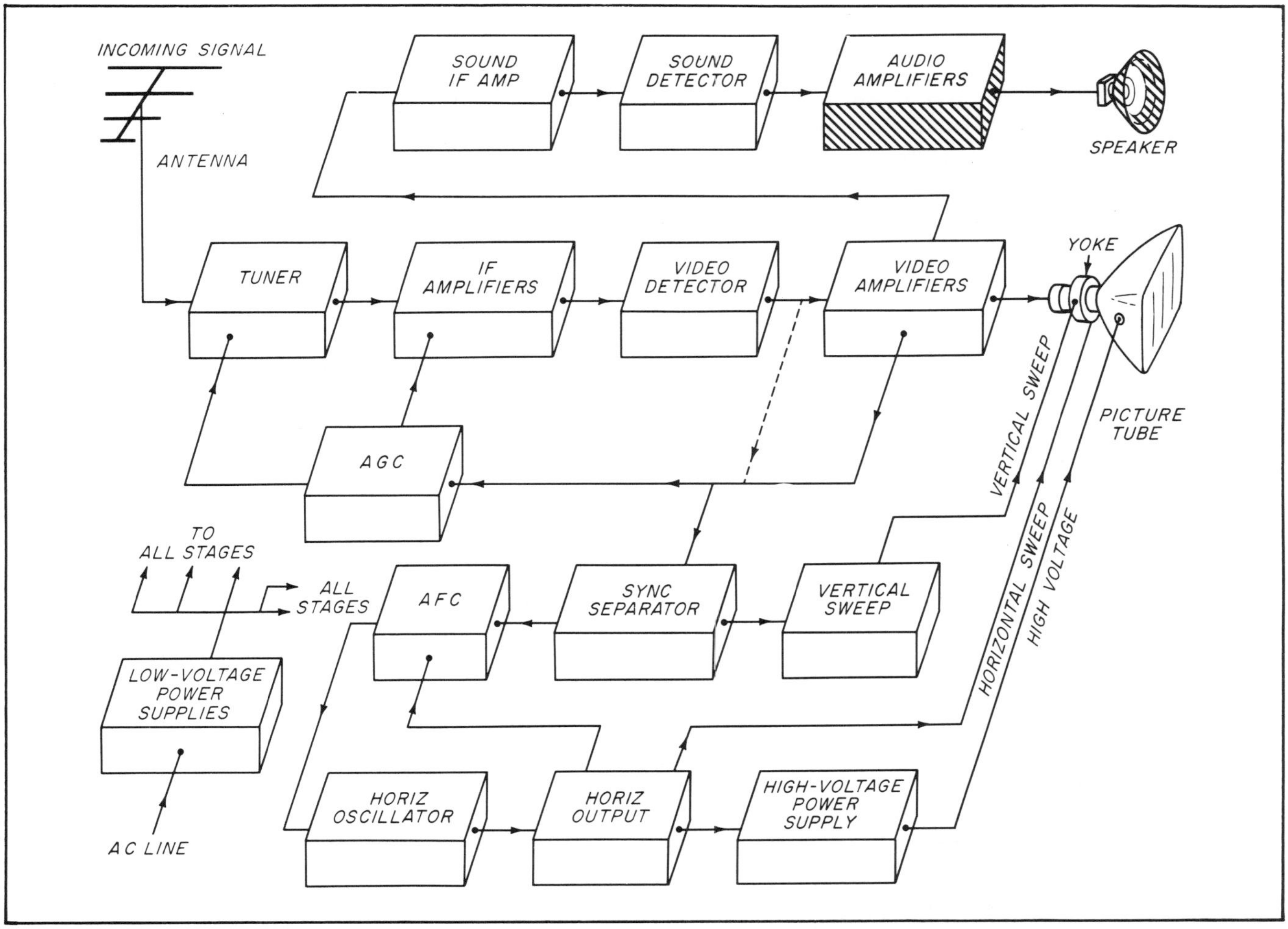

Fig. 3-1. Block diagram of a typical black-and-white receiver.

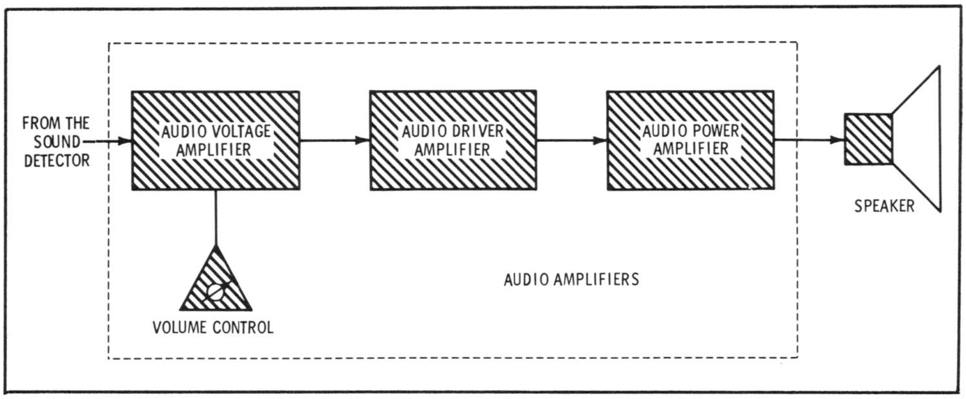

Fig. 3-2. Audio section of tv receiver.

the ones that affect *only* the audio performance. Within the audio-amplifiers block there may be two or three different amplifier stages.

Fig. 3-2 shows a typical audio-section arrangement. An important feature to notice is that the *volume control* is located in the audio voltage-amplifier stage. This fact can be used to help identify audio-circuit problems.

In diagnosing audio-section troubles, listen very carefully to the speaker and vary the volume-control setting. If the audio section is working normally, the sound should go up and down with the volume control. At maximum volume, the sound from the speaker should be quite loud and undistorted. At minimum volume, only a very soft humming sound should be heard. *Most* variations from these characteristics can be considered as audio trouble symptoms. The most common symptoms are:

1. Little or no sound at maximum volume
2. Squealing or whistling sound
3. Distorted sound output

Each of these three major sound-trouble symptoms will be considered in some detail in this chapter. In addition to the three symptoms listed above, it is possible that the sound may also contain a severe humming or buzzing noise. These latter symptoms usually indicate trouble in *another* part of the set.

Assume the situation of a set which has no sound output. The test of varying the volume control and listening carefully to the speaker can still be used. If the audio circuits are working normally, a very soft hissing noise in the speaker will be heard. This hissing noise should go up and down a little as the volume control is turned.

If there is a failure within the audio system, then varying the volume control usually will not change the noise level at the speaker. When this is the case, listen very carefully for a soft humming sound. This soft hum means that the speaker and audio power amplifier are probably functioning normally, and any audio-section failure must be in the voltage amplifier or driver amplifier. On the other hand, if there is *no* soft hum, it means that there is a failure in the power amplifier or speaker.

Squealing, whistling, or distortion in the sound are also important audio symptoms. These things will be discussed further after we look at some audio-circuit details.

3-2 Electric Circuits

Most people do not understand electricity very well even though it is all around them. In this section, the way electricity works will be examined. It is not at all hard to understand; in fact, electricity is very much like water in many ways. It flows around a *circuit* similar to the flow of water through a pipe. Suppose you have a water pump and some pipes set up as shown in Fig. 3-3.

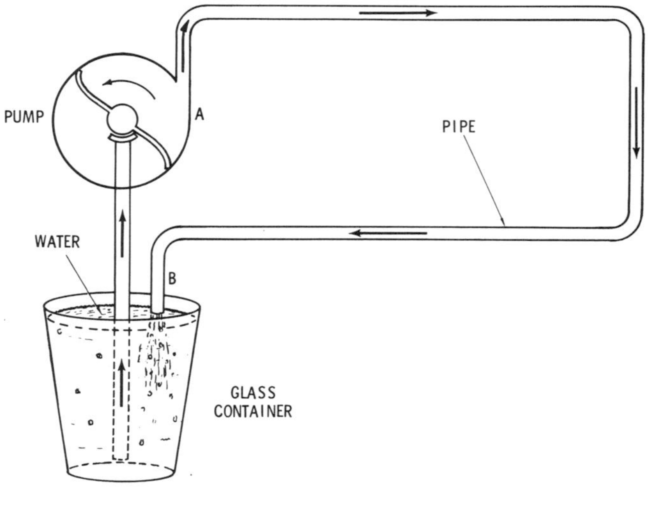

Fig. 3-3. Water flow around a circuit.

If the pump is turned, it will pick up some of the water from the glass container and push it into the pipes at point A. Then if the pipes are all full, this new water will push all of the water on around the pipes. When this happens, some water will run back into the glass at point B. The faster the pump is turned, the more water flows around the circuit.

In an electric circuit, the pump and glass could be a battery, and the pipes could be wires. The only difference is that the battery cannot be told how fast to pump.

If the pump in Fig. 3-3 turned at the same speed all the time, it would be just like a battery. The pressure at the output of the pump could be measured, and if it turned out to be 9 pounds per square inch, we could call this a 9–pound-per-square-inch pump. This is similar to what is done in electricity, but a battery's pressure is measured in *volts,* not pounds per square inch, so you could have a 9-volt battery.

In the water circuit, if a section of the pipe is replaced with very small pipe, less water will flow. In electricity, if a big wire is replaced by a very small one a similar thing happens—less electricity flows, and we say the *current* is less. It could be said in either case that the smaller pipe (wire) had a high *resistance* to flow.

These three things (voltage, current, and resistance) are really the basic elements of all electric systems. What an electronic designer does is to select the amount of voltage and resistance which will give enough current to do a particular job.

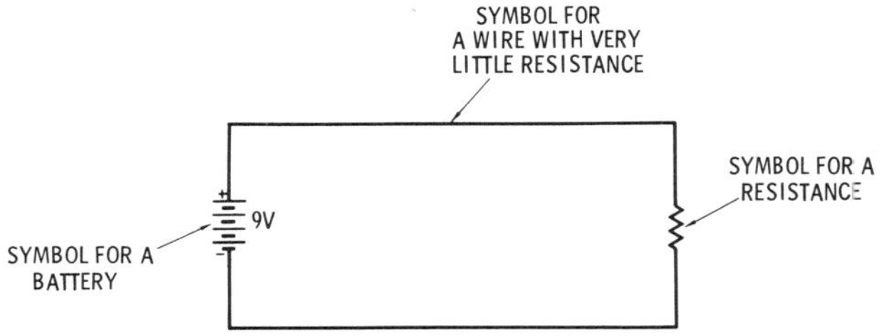

Fig. 3-4. An electric circuit.

When an electric circuit is drawn, *symbols* are used to represent these basic elements. Fig. 3-4 shows the electric circuit that is equivalent to the water circuit of Fig. 3-3. In electricity, the amount of resistance a circuit has is given in terms of *ohms.* That is, it could be said that a given circuit has 1000 ohms of resistance. The more resistance a circuit has, the less will be the current in the circuit. Current is measured in amperes (amps) or milliamperes (one milliamp is one thousandth of an amp) or microamperes (one microamp is one millionth of an amp). Milliampere is abbreviated mA, and microampere is abbreviated μA.

Complicated circuits can be made by using several resistances together. A *series circuit* is one in which resistors are hooked end-to-end as shown in Fig. 3-5A. A *parallel circuit* is one in which the resistors are hooked up side-by-side (Fig. 3-5B). If series and parallel circuits are connected together, the combination is called a series-parallel circuit (Figs. 3-5C and 3-5D).

In an electronic circuit like a television receiver, two kinds of electricity are found. Direct-current (dc) electricity comes from a battery or a dc power supply. The voltage level of dc electricity does not change from time to time while it is being used. Alternating-current (ac) electricity changes its voltage level up and down all the time it is being used. Ac electricity is available at the wall outlet, and this ac is used to power the set. At a wall outlet, the ac is about 120 volts and goes up and down (alternates) 60 times every second. We would describe it as being 120 volts, 60 hertz (Hz). Inside a set, the dc power supplies convert this 120-volt, 60-Hz ac into dc to operate the electronic circuits. The tv signals from the station are also ac voltages, but they alternate many millions of times each second.

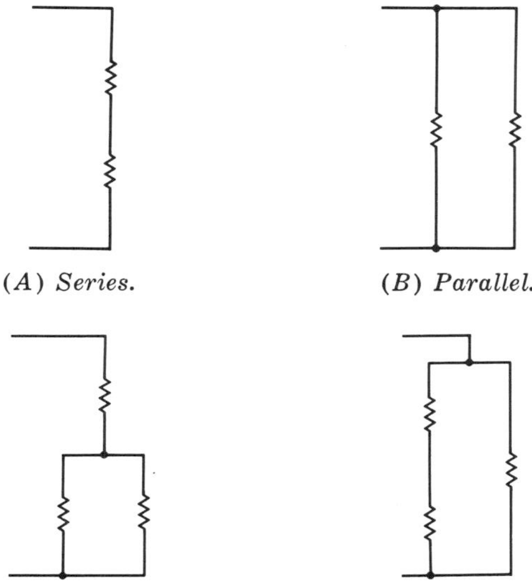

(*A*) *Series.* (*B*) *Parallel.*

(*C*) *Series-parallel.* (*D*) *Series-parallel.*

Fig. 3-5. Various combinations of resistances.

In addition to resistors, there are three other types of electric circuit components found in a television set. They are:

1. Capacitors, which pass ac but block dc electricity

2. Inductors, or coils, which are used to block ac and pass dc electricity
3. Transformers, which can step ac up or down (raise or lower the voltage)

Fig. 3-6 shows the circuit-diagram symbols used for these components.

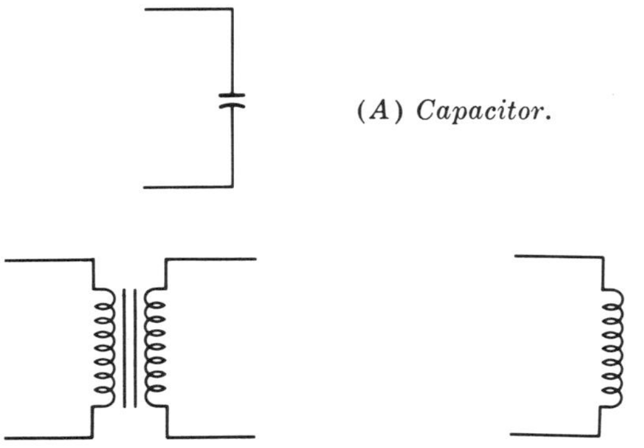

(*A*) *Capacitor.*

(*B*) *Transformer.* (*C*) *Inductor.*

Fig. 3-6. Symbols for common components.

3-3 Speaker Circuits

The speaker is the last component in the audio section. If the speaker circuit fails, there will be no sound and no soft hum present.

The speaker is constructed of a paper *cone* with a coil of wire wrapped around a cylinder near the apex as shown in Fig. 3-7. The coil of wire is called the *voice coil.* It is suspended in the field of a magnet. When an audio signal is passed through the voice coil, it acts like a motor and moves the cone. The cone moves the air, thus producing sound waves that can be heard.

If the cone or voice coil gets stuck, the speaker will not work. The voice coil can be checked by gently pushing on it with the fingers. It should move, and the smooth springiness of the cone should be felt.

If the voice coil has a broken wire, the speaker will not work. The voice coil can be checked by measuring its resistance with an ohmmeter. To do this, disconnect the speaker from its circuit, and connect one ohmmeter wire to one speaker terminal. Set the ohmmeter to its lowest resistance range, and touch the remaining ohmmeter wire to the remaining speaker terminal. You should get two noticeable results. First, you should *hear* a single click from the cone as the meter is connected. And, you should *see* a meter reading of very few ohms. If both results are present, the speaker will probably work all right.

The speaker circuit of a typical television receiver is shown in Fig. 3-8. Notice that an earphone jack, speaker plug, and the audio output transformer are also parts of the speaker circuit. After the speaker is checked, these other components should also be tested. A good practice is to unplug the speaker plug and repeat the speaker check at the plug. If both results described previously (the click and low ohms reading) are present, then the speaker and earphone plug are all right. The output transformer winding may be tested with the ohmmeter while the speaker is unplugged. It should measure about the same resistance as the speaker voice coil.

3-4 The Audio Power-Amplifier Circuit

More than anything else, electronics is the study of devices that amplify. Probably the most com-

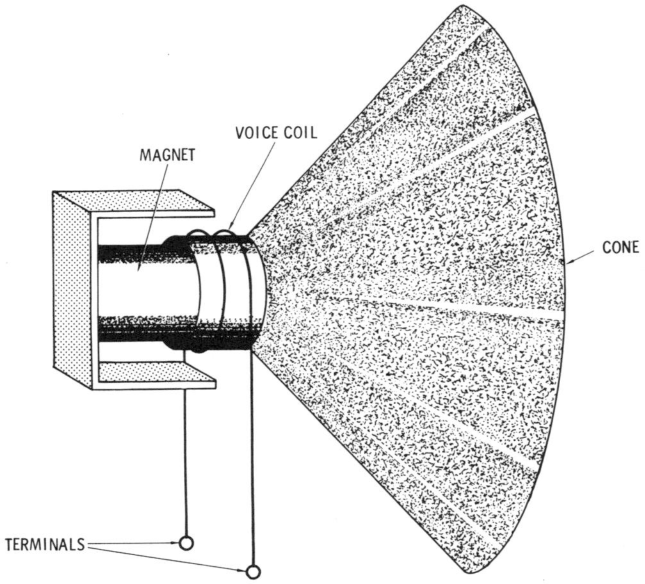

Fig. 3-7. Main parts of a speaker.

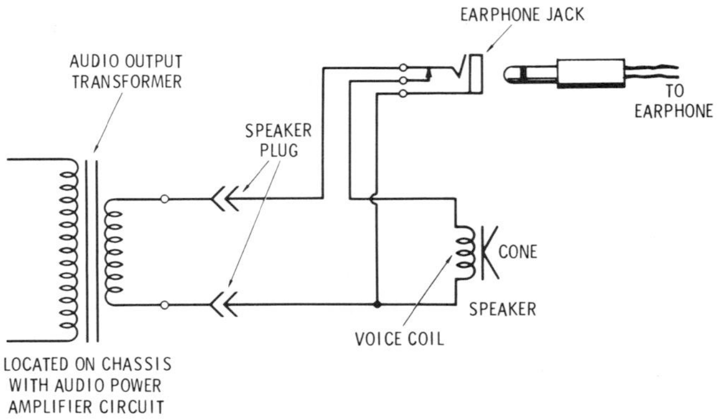

Fig. 3-8. A typical speaker circuit.

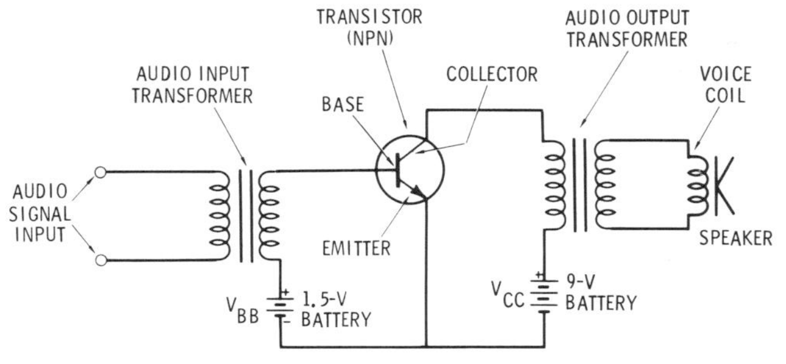

Fig. 3-9. A basic amplifier circuit.

mon such device is the ordinary transistor. A transistor is nothing more than a device whose resistance can be controlled by a very small current. Fig. 3-9 shows a simple transistor amplifier circuit.

In this circuit, there are two different paths that current can follow. One of them is from the 9-volt battery over to the *emitter terminal* of the transistor, up through the transistor to the *collector terminal,* then through the audio output transformer back to the 9-volt battery. This is called the collector circuit. The current through this path is called the *collector current.* The 9-volt battery is called the *collector supply* and is often a dc power supply rather than an actual battery. The size of the collector current is normally several milliamperes. Since it is dc, it is not coupled through the output transformer to the speaker.

On the other side of the transistor, current can pass from the 1.5-volt battery to the emitter, and through the transistor to the base connection. From there it flows through the input transformer and back to the 1.5-volt battery. This base current is normally only a few microamperes. Notice that the collector current is many times greater than the base current.

Now, if an audio ac signal is connected to the audio input transformer, the signal is coupled into the base circuit. This causes the base current to vary in step with the audio signal. The variation in base current causes the collector current to vary in step with the audio signal. As the collector current increases, the voltage drop across the primary of the audio output transformer increases. The collector voltage must decrease as the drop across the transformer increases. Therefore, as the base voltage rises, the collector voltage drops. We describe this inverse relationship by saying that *there is a 180-degree phase shift from the transistor base to its collector.*

The collector current varies by about the same percentage as does the base current. However, since the collector current is much greater than the base current, the transistor can be said to have *amplified* the base-current variations. For example, a base-current change of 5 microamperes can cause a collector-current change of 1 milliampere.

This variation in collector current, being an ac variation, is coupled through the audio output transformer to the speaker. At the speaker, the current variations are converted to audible sound.

While the circuit shown in Fig. 3-9 serves very well to demonstrate power-amplifier operation, it is not really a practical circuit. Fig. 3-10 shows a typical audio power-amplifier circuit. The operation of this circuit is the same as described before, except that the base current is now supplied by the same dc power source that supplies the collector current. The 18,000-ohm and 10,000-ohm resistors determine the amount of dc bias that is supplied to the base. These resistors are called *base-bias resistors.*

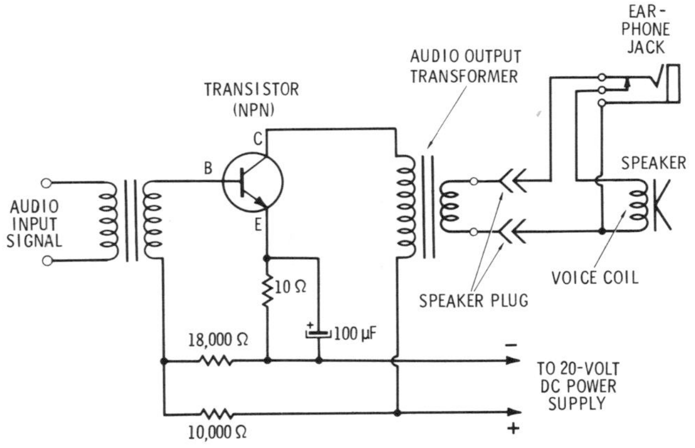

Fig. 3-10. A typical power amplifier.

Also, a resistor and capacitor have been added to the emitter circuit. The *emitter resistor* helps protect the transistor against excessive collector current. The *emitter capacitor* prevents the emitter resistor from interfering with audio signal operation.

The transistors shown in Figs. 3-9 and 3-10 are of the *npn* type. There is another type called a *pnp* that is also used in power amplifiers. The operation of both is the same. However, the dc power-supply connections are reversed. The negative power-supply terminal is connected to the collector circuit of a pnp transistor. The positive terminal is connected to the emitter in a pnp circuit. The electrolytic-capacitor leads must also be reversed. The symbol for a pnp transistor is identical to an npn one except that the arrowhead on the emitter is reversed.

3-5 Troubleshooting a Power Amplifier

If there is no sound and no soft hum from a tv set, then the speaker or power-amplifier circuit should be suspected. If the speaker circuit checks out all right, then the power amplifier should be checked.

The first step in checking any electronic circuit is to measure the dc supply voltage at the circuit. Suppose that the supply voltage in Fig. 3-10 reads exactly 20 volts. Next check the voltage from collector to emitter of the transistor. In a power amplifier, the collector-to-emitter voltage should be a volt or two less than the supply voltage. If it is too high or too low, check the transistor.

A power transistor can be tested by removing it from the circuit and measuring the terminal resistances with an ohmmeter. To do this, it must first be determined which ohmmeter wire is positive and which is negative. Then connect the negative ohmmeter lead wire to the base, the positive ohmmeter lead to the collector, and read the resistance. It should be very high for an npn transistor and very low for a pnp transistor. Now move the positive lead to the emitter and repeat the reading. The results should be the same. Reverse the positive and negative ohmmeter wires and repeat the process. The results should reverse. In summary form the results should be as shown in Table 3-1.

Table 3-1. Transistor Tests

NPN Transistors			
Negative Base		Positive Base	
High Collector R	High Emitter R	Low Collector R	Low Emitter R
PNP Transistors			
Negative Base		Positive Base	
Low Collector R	Low Emitter R	High Collector R	High Emitter R

Any severe variation from this pattern indicates a defective transistor.

The transformer windings in a power amplifier may be checked using an ohmmeter, and so can the resistors. The capacitor in Fig. 3-10 can be checked with an ohmmeter to see if it is short-circuited.

It is a good practice to replace a suspected part temporarily with one known to be good. This can only be done, of course, if known good parts are available.

A quick overall check of a power amplifier is to connect a voltmeter across the emitter resistor. It should read about 1 volt. Then *momentarily* short-circuit the bias resistor that has one end connected to the emitter (18,000 ohms in Fig. 3-10). The voltage should drop to zero and there should be a "click" in the speaker. Do not leave the resistor shorted more than a few seconds. *Be very careful not to short the base to the collector.* This could damage the transistor.

3-6 Push-Pull Power Amplifiers

Some television receivers use two power-amplifier transistors to operate the speaker. Fig. 3-11 shows this type of audio power-amplifier circuit. This type of circuit is called a *push-pull amplifier.* The operation of a push-pull type amplifier is just like a single-transistor type, except that when one collector current is decreasing, the other is increasing. In this way, it is possible to produce a somewhat more efficient amplifier.

Troubleshooting a push-pull amplifier is about the same as with a single-transistor type. However, the quick check used with a single-transistor circuit does not work well with a push-pull circuit. A much more effective check is to insert an audio test signal from a signal generator at the input and listen to the output.

3-7 Audio Voltage-Amplifier Circuitry

If there is no sound from the speaker but a soft hum is present, the tendency is to think that the speaker and power amplifier are all right. This being the case, vary the volume control and listen for any change in noise. If the noise level does not change, a failure should be suspected in the audio voltage amplifier or driver circuit.

Fig. 3-12 shows a typical audio voltage-amplifier stage. While the amplifier works in basically the same way as the one discussed previously, there are a few differences in circuitry.

The audio signal from the sound detector stage is applied directly across the volume control (R1). The volume-control center contact moves up and down along R1 as the volume-control knob is turned. The center contact picks off a part of the audio signal and couples it to the transistor base

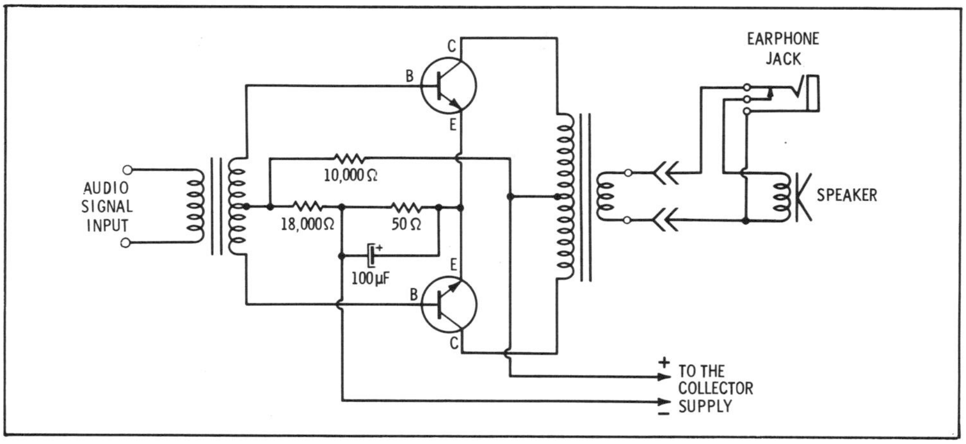

Fig. 3-11. A push-pull power amplifier.

through capacitor C1. This capacitor is called the *input coupling capacitor*. It prevents dc from passing between the detector and amplifier circuits. The base-bias resistors (R2 and R3) serve as before to establish the dc base current. The emitter resistor serves to stabilize the transistor against internal temperature variations; this prevents damage due to excessive collector current. The emitter bypass capacitor (C3) prevents the emitter resistor from interfering with the circuit's ac operation. The collector circuit contains the collector load resistor (R5) and the output coupling capacitor (C2). When the collector current varies with the audio input signal, these variations are coupled out of the stage through the coupling capacitor. The collector resistor assists in establishing the desired dc collector current.

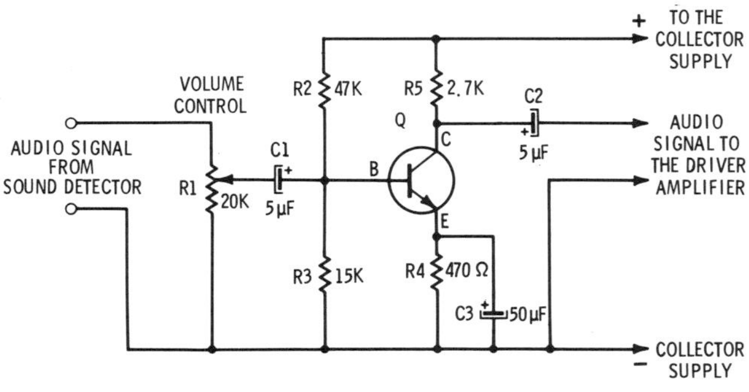

Fig. 3-12. An audio amplifier circuit.

Notice that the resistor values shown in this circuit have a K or an Ω (omega) with them. Used in this way, Ω stands for ohms and K stands for thousands of ohms. Therefore 470 Ω means 470 ohms and 47K means 47 thousand ohms. Capacitor values are shown in microfarads (1/1,000,000 farad). Microfarad is abbreviated µF (sometimes mfd). Another unit of capacitance used is the picofarad, which is 1/1,000,000 of a microfarad and is abbreviated pF (sometimes pf or µµF).

In troubleshooting an audio amplifier of this type, proceed as with a power amplifier; that is, first check the collector supply voltage. Then check the collector-to-emitter voltage at the transistor. In this kind of circuit, the collector-to-emitter voltage is expected to be slightly over one-half the supply voltage. For example, if the supply voltage is 20 volts, then the collector-to-emitter voltage should be 11 or 12 volts. It it is considerably different from this, a bad transistor or other component should be suspected. Components are checked using an ohmmeter.

The transistor shown in Fig. 3-12 is an npn type. Audio amplifiers are often built using pnp transistors. The only differences are:

1. The supply voltage is reversed.
2. The capacitors are all reversed.
3. The pnp transistor symbol is used.

Otherwise, the circuits are the same.

Audio driver amplifiers have circuit diagrams that are almost identical to audio voltage amplifiers.

The circuit diagrams supplied for the television set usually show the dc operating voltages to be expected at various points in the circuit. This information is very helpful in troubleshooting a receiver circuit.

3-8 Troubleshooting the Entire Audio Section

Fig. 3-13 shows an audio section from a television receiver. Let us discuss how the circuit operates first and then consider how to troubleshoot it.

The audio signal from the sound-detector circuit is applied to the volume control. From the volume control the signal is coupled to the base of the

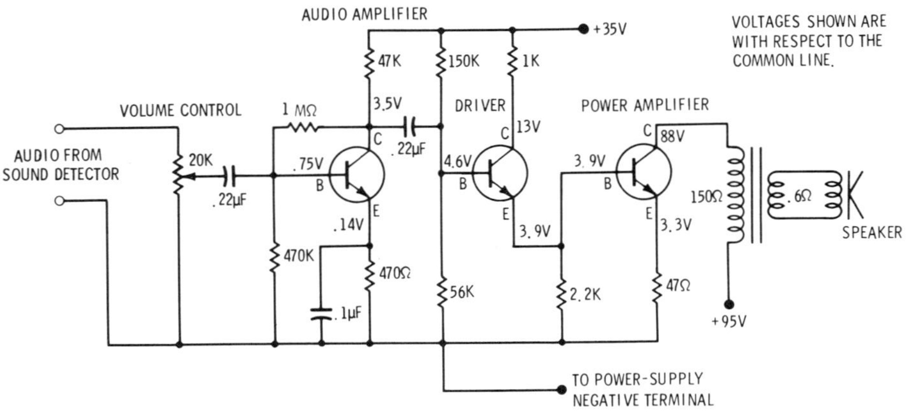

(*A*) *Complete section with output transformer.*

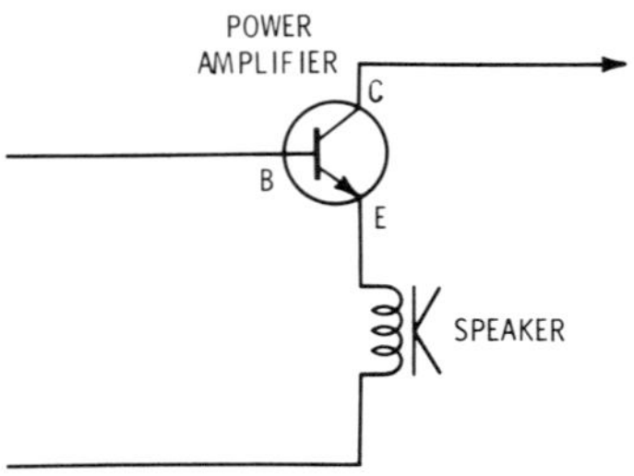

(*B*) *Speaker circuit with direct coupling.*

Fig. 3-13. A typical audio section.

audio amplifier. The audio-amplifier circuit is just like the one discussed previously except that the 1-MΩ (one million ohms) resistor is connected to the collector instead of to the power-supply line. This type of circuit gives a more constant but lower amplification than the one discussed before.

From the audio amplifier, the signal is coupled to the driver amplifier through a 0.22-μF capacitor. This driver circuit has the output taken off at the emitter instead of at the collector. It works like any other amplifier but it has two interesting characteristics. They are:

1. The power amplification is between that of an audio amplifier and a power amplifier, but the voltage amplification is very small.
2. There is no phase shift. That is, the output goes up when the input does and vice versa.

Notice that the emitter voltage of the driver is the base bias for the power amplifier. This type of coupling is called *direct coupling* and is used quite a lot in television circuitry. Other than the direct coupling, the power amplifier is just like the one we have already discussed.

In addition to direct coupling between stages, the speaker may also be direct coupled, as shown in Fig. 3-13B. In such a case, care must be exercised not to short-circuit or remove the speaker connections. Either action will change the voltages throughout the three amplifier stages and *can result in damage to the circuit.*

In troubleshooting an audio section, the problem must first be sectionalized by carefully listening to the speaker while varying the volume control up and down. If the speaker noise does not vary with the volume control, then the trouble is in the audio section.

Next, localize the trouble by listening for a soft hum from the speaker. If there is no hum, the speaker itself or the power amplifier is suspected. The speaker and output transformer can be checked with an ohmmeter after disconnecting them from the circuit.

The voltages shown on the diagram of the power amplifier can be checked with a voltmeter. If the transistor voltages are very different from those shown, remove the transistor and test it with an ohmmeter or transistor checker.

It is worth noting at this point that since the driver and power amplifier are direct coupled, any change in driver voltage is coupled directly to the power amplifier. Consequently, the driver voltages should be checked at this time.

The resistors in the power-amplifier and driver circuit can be checked with an ohmmeter. Any uncertain part may be temporarily replaced with a known good one. This test procedure will locate any defective part in the power-amplifier circuit.

If there is a soft hum in the speaker but the speaker noise does not vary with the volume control, the audio amplifier or driver should be suspected.

Always start testing at the driver collector and move back toward the volume control. If the driver circuit voltages are normal, then move back to the audio amplifier circuit. The audio amplifier is tested by measuring the transistor voltages first. If they are not correct, the circuit components are tested one at a time.

3-9 Integrated-Circuit Amplifiers

Many modern television receivers have what is called an *integrated circuit* in the audio section. An integrated circuit (IC) is nothing more than one or more amplifier stages sealed up in a small package. Fig. 3-14 is a typical diagram of an audio section that contains an IC.

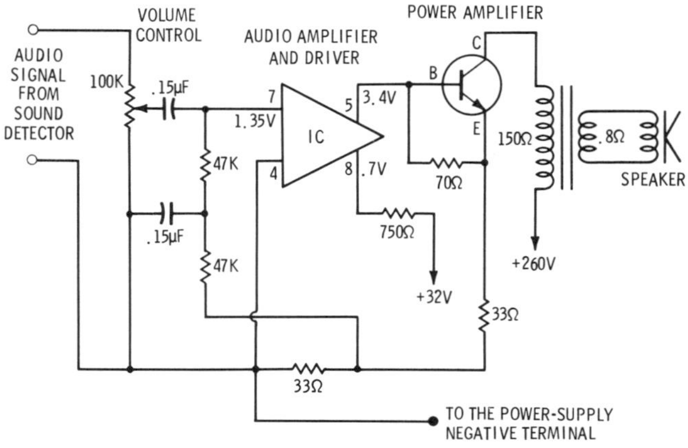

Fig. 3-14. An IC audio section.

Some sets use an IC circuit that has the power amplifier inside the integrated package. When this is the case, the transistor in Fig. 3-14 will be omitted, and the speaker will connect directly to the IC output.

Another common circuit variation is to use the volume control to adjust the bias on the IC rather than controlling the audio signal. When this is done, the audio is coupled straight into the IC, and the volume control connects to other terminals on the IC.

In troubleshooting an IC system, proceed just as before. However, since the individual components inside the IC cannot be tested, it is only necessary to measure its terminal voltages. If they are very different from the values shown on the manufacturer's diagram, and all other parts test all right, then replace the whole IC.

3-10 Distortion, Squeals, and Whistles

Occasionally, a television receiver will have sound that is distorted or has squealing, whistling, or howling noises mixed with it. Some of these symptoms can be caused by other than audio-section failures. However, they are *usually* caused by a bad audio transistor or a bad capacitor in an audio stage. There are normally only 6 or 8 of these components. Therefore, a good solution is to replace them temporarily, one at a time, with known-good units until the defective part is found.

3-11 Vacuum-Tube Audio Amplifiers

Many television receivers use vacuum tubes in the audio section. There is not much difference in the way troubleshooting is carried out. Fig. 3-15 shows a vacuum-tube audio section. There will normally be only one or two stages in the section. In locating a defective component, the tube is usually replaced first. If this does not cure the problem, then the terminal voltages are checked. The defective component is usually one of those connected to the terminal with incorrect voltage.

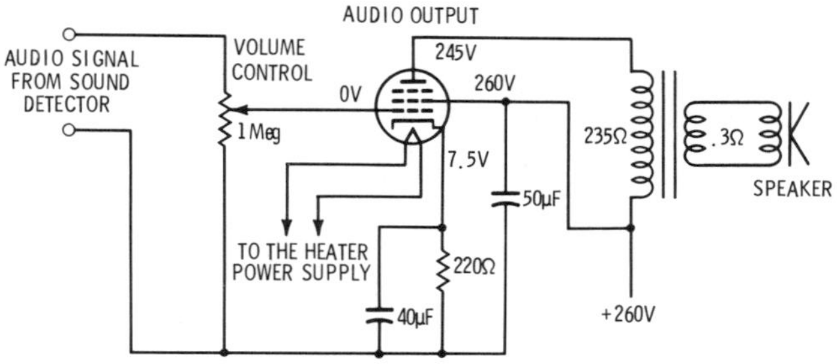

Fig. 3-15. A vacuum-tube audio section.

The circuit diagrams of several sets are included at the back of this book. You should examine the audio-section circuity of each one. This will help you become familiar with typical circuitry.

3-12 Review Questions

1. What are the symptoms of a defective speaker?
2. How do you test a speaker?
3. If the speaker of a set is tested and found to be good, what do you do next?
4. What is an IC?
5. How do you test an integrated circuit?
6. A television set has no sound, but a soft hum is present at the speaker. Where in the set would you look for the problem?
7. The audio section illustrated in Fig. 3-13 is being tested. The +35 V measures normal, but the collector voltage of the audio amplifier is zero. Which components could cause these symptoms?

8. In Fig. 3-13, the base voltage of the power amplifier is 10 volts. Which components would you check?
9. The test point marked 245 V in Fig. 3-15 measures zero. Which components would you suspect of being defective?
10. What are the normal symptoms of a dead audio amplifier?
11. How do you distinguish between a defective power amplifier and a defective voltage amplifier?
12. When the volume control is varied, the background noise varies, but there is no sound. What do these symptoms suggest?
13. The voltage at pin 5 of the IC in Fig. 3-14 measures zero. What would you do next?
14. Explain why the driver in Fig. 3-13 could be considered part of the power amplifier.
15. The audio section in Fig. 3-13 produces severely distorted sound. Which components would you suspect?
16. How would you test a pnp transistor?
17. How would you test a volume control that you thought to be defective?
18. Draw a circuit diagram like Fig. 3-10 using a pnp transistor.
19. What is a voice coil? Where is it found in a television receiver?

CHAPTER 4

The Sound IF Section

Suppose that you have a set that does not have any sound output. In diagnosing the problem, first look at the picture. If the picture is normal but the sound does not work, the problem must be in the sound circuits somewhere.

The sound section can be thought of as having two parts, the audio stages and the sound if stages. If there is a failure in one of the audio stages, it can usually be determined by varying the volume control. Listen very carefully to the speaker and run the volume control up and down. If there is no change at all in the level of the soft hum and

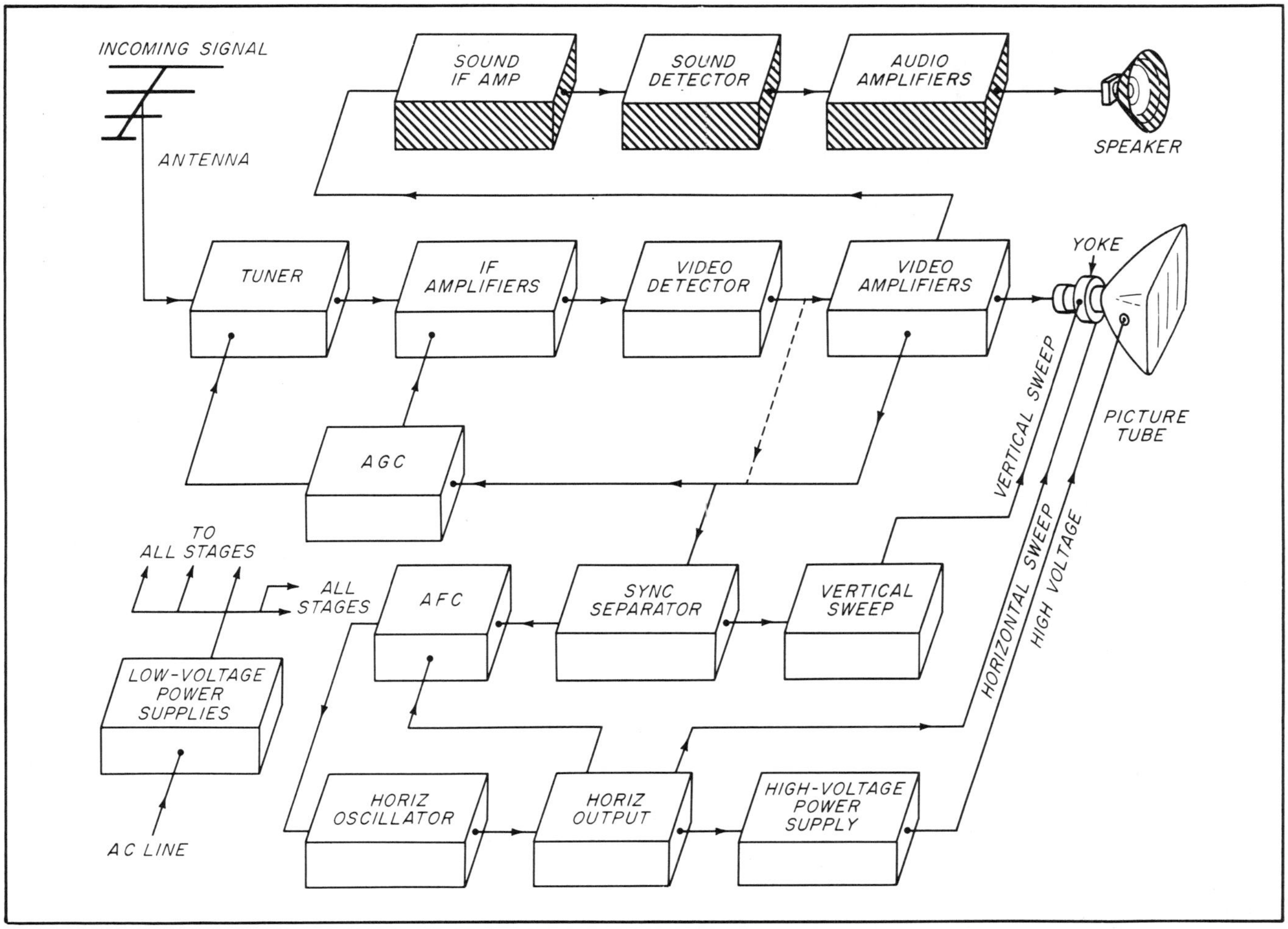

Fig. 4-1. Block diagram of typical black-and-white receiver.

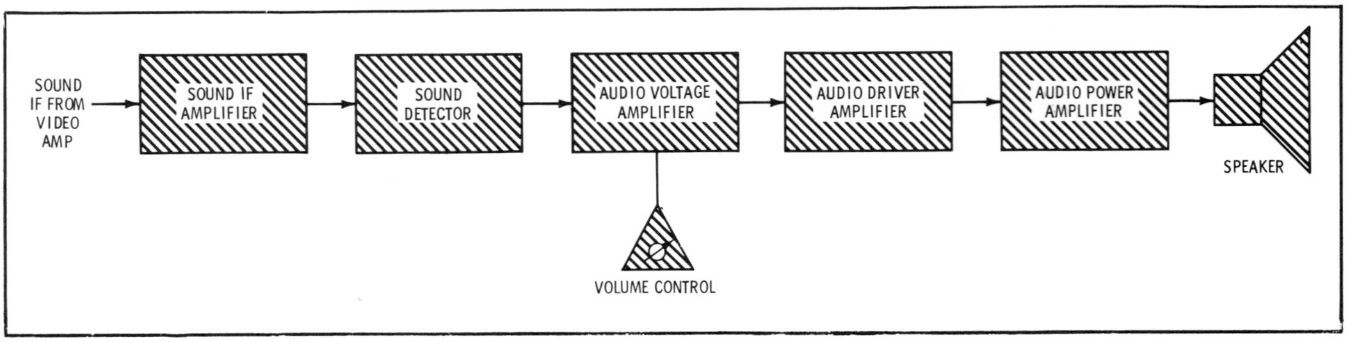

Fig. 4-2. The television sound system.

background noise, then the falure must be in the audio circuits. On the other hand, if there is a little change in the background noise, then the audio stages are probably all right.

4-1 Identifying the Trouble

Now, what does this indicate so far? The picture on the set looks normal and the audio amplifiers are working, but the sound is dead.

Look at the block diagram shown in Fig. 4-1. It can be seen that the sound comes from the video amplifier. But since the picture is normal, the video amplifiers are working. From the video amplifier, the sound goes through the sound if and sound-detector blocks before it gets to the audio stages. If there is a problem in either the sound if amplifier or the sound detector, the symptoms described above will appear.

In summary then, if there is no sound from the set, but a little change in background noise can be heard when the volume control is varied, a problem in the sound if or detector stage should be suspected.

4-2 Sound Section Operation

Before looking at the sound if amplifier and detector circuits, pause and go over the way the whole sound section works. Fig. 4-2 shows a detailed block diagram of the sound section.

The sound signal, which comes from the video amplifier, is an fm radio signal at a frequency of 4.5 MHz (4.5 megahertz, or 4.5 million cycles per second). The sound section in a tv set is really a simple fm radio. Since the sound signal is always 4.5 MHz, this fm radio does not have to be tuned from station to station; it is adjusted at the factory to respond to that one frequency.

After the fm radio signal is separated from the video, it goes to the sound if amplifier. In this

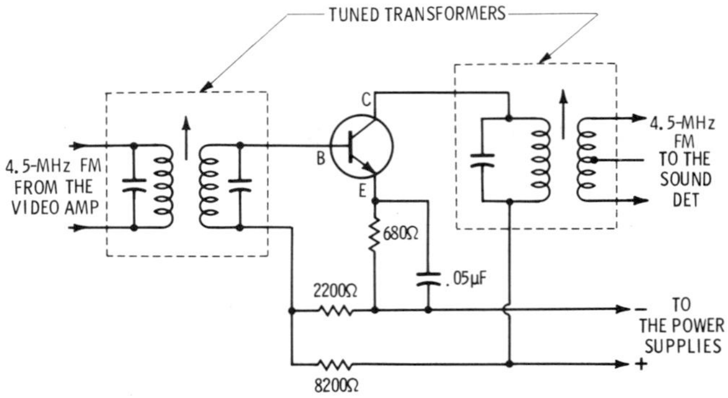

Fig. 4-3. A sound if amplifier.

stage, the signal is simply amplified. The signal then goes to the detector, where the fm radio signal is converted to an audio signal. This audio signal is sent through the audio stages to the speaker.

4-3 The Sound IF Amplifier

The sound if amplifier is a *fixed-tuned radio-frequency amplifier*. That is, it amplifies only one frequency. Fig. 4-3 shows a simplified sound if amplifier circuit.

The if amplifier works just like an audio amplifier except for one thing. The tuned if transformers are set so that the amplifier will accept only signals that are very close to 4.5 MHz. As a

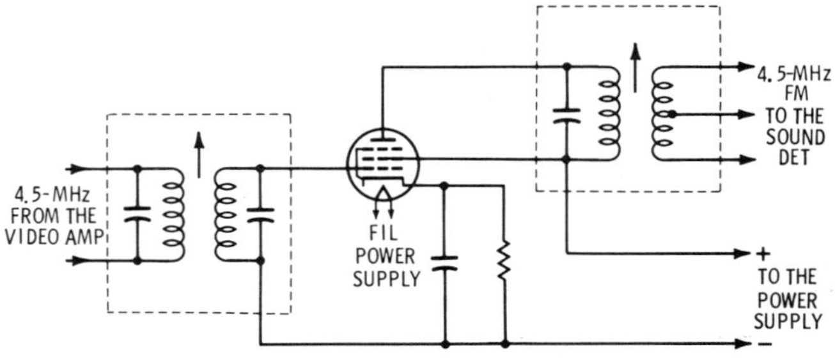

Fig. 4-4. A vacuum-tube if amplifier.

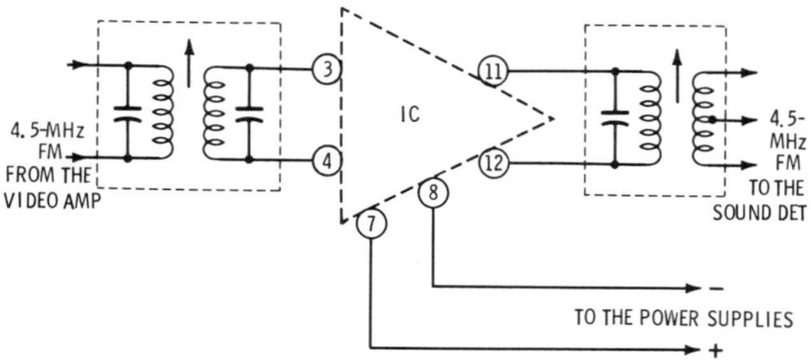

Fig. 4-5. An integrated-circuit if amplifier.

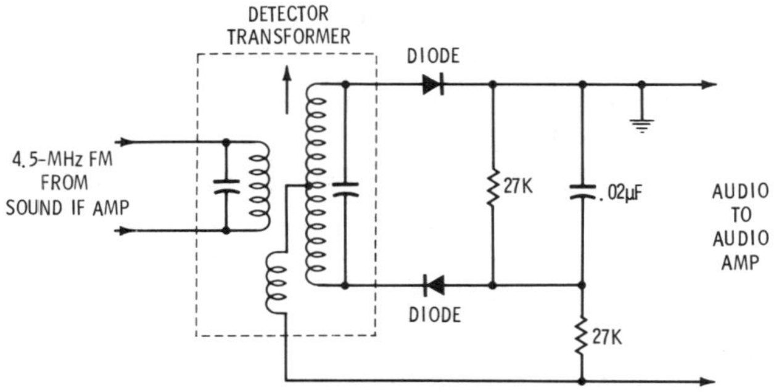

Fig. 4-6. A television sound detector.

result, the circuit amplifies only that one band of frequencies.

In some sets, the sound if amplifier may use a vacuum-tube circuit like the one shown in Fig. 4-4. But again, the operation is the same as in any other amplifier except for the tuning. In most hybrid sets, the sound if amplifier stage is in a tube that has several sections. One single tube may contain the sound if amplifier and perhaps the audio amplifier as well as the audio power amplifier.

Integrated circuits are also sometimes used for sound if amplifier circuits. Fig. 4-5 shows a circuit diagram of this type.

As is the case with a multisection tube, many integrated circuits have several sections. It is not uncommon to find a single IC that contains the sound if amplifier, the sound detector, and an audio amplifier.

4-4 The Sound-Detector Stage

The signal coming from the sound if amplifier is a 4.5-MHz fm radio signal. This radio signal varies in frequency at the sound rate. The job of the sound detector is to convert the frequency variations of the fm signal into an audio signal. Fig. 4-6 shows a typical sound-detector circuit.

The sound-detector circuit is a very simple one and has proven to be quite reliable. It is very rare

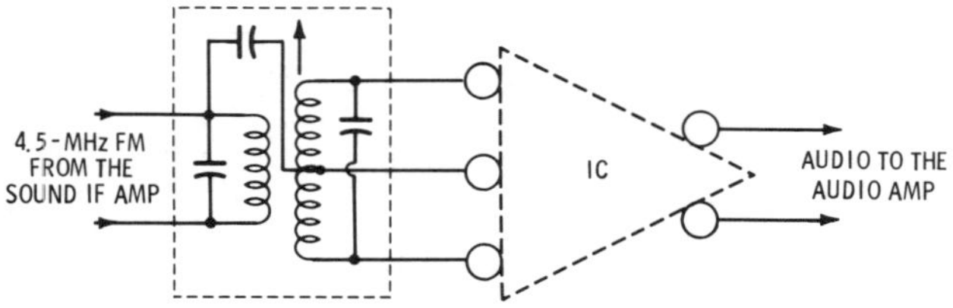

Fig. 4-7. An IC sound detector.

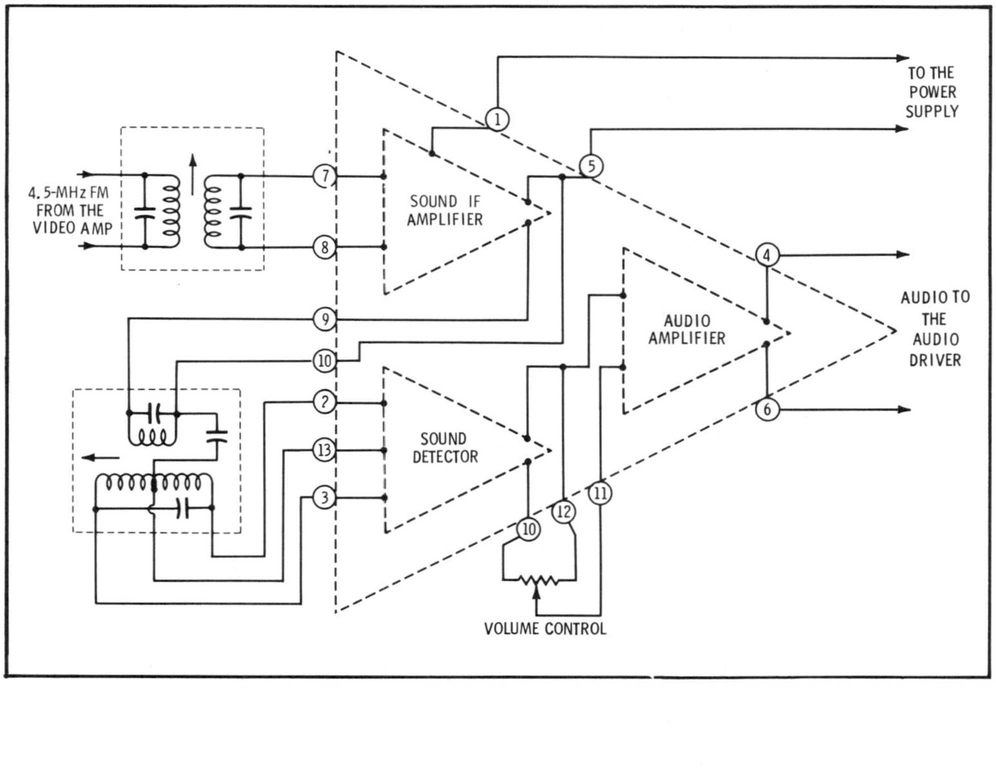

Fig. 4-8. An integrated sound circuit.

for the detector circuit to fail. As a result, the sound if amplifier is usually suspected more than the detector. One problem that does arise occasionally is a buzz in the sound caused by an improperly tuned detector. Some sets even have a buzz control built into the detector circuit.

Vacuum tubes are not often used in sound-detector circuits, but integrated circuits are. Fig. 4-7 shows a typical IC detector circuit.

It is common for the detector and sound if amplifier to be integrated in the same IC package. Fig. 4-8 shows a circuit in which both stages are in the same "chip." The illustration also shows an audio amplifier in the IC with the sound if and detector. In an integrated sound circuit of this type, the whole IC will have to be replaced if any one of the three circuits fails. With discrete circuitry (such as that shown in Fig. 4-9), only the defective component would require replacement.

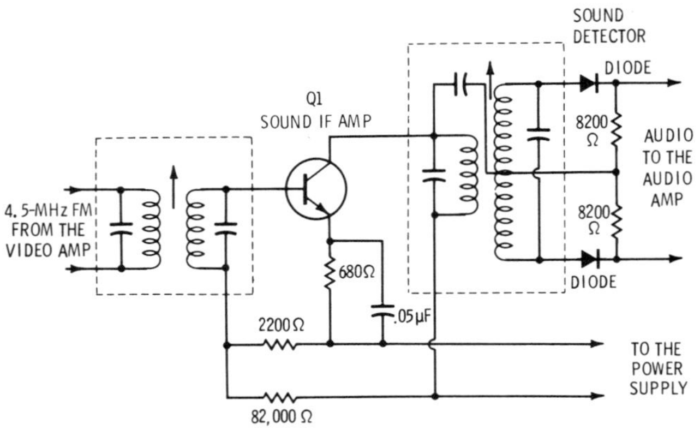

Fig. 4-9. A solid-state sound if section.

4-5 Checking the Sound IF Circuits

When a failure in the sound if amplifier or detector is suspected, measure the circuit voltage. In the case of a transistor set, the collector-to-emitter voltage of the if amplifier should be measured. It should be about 90 percent of the collector supply voltage. The emitter-to-base voltage is usually less than about 10 percent of the collector supply voltage—typically 0.1 to 0.8 volt.

With an integrated circuit, the best thing to do is to measure the IC terminal voltages and compare them to the values shown on the circuit diagram of the set.

The voltage from plate to cathode of a vacuum-tube if amplifier usually runs about 95 percent of the plate supply value. The screen-grid voltage should be about the same as the plate voltage. The control-grid–to–cathode voltage is usually about 5 percent of the plate supply voltage, and the grid is negative with respect to the cathode.

Whenever possible, the circuit diagram of the set you are working on should be obtained. Then the measurements can be checked against values for normal operation. In doing this, remember that there are normally small differences between measured values from set to set. Do not expect the measured values to agree *exactly* with those on the diagram.

4-6 Summary of Sound IF Symptoms

When the sound if amplifier or sound detector fails, the picture will remain normal. The sound will be dead, but some variation in the background noise can usually be heard when the volume control is varied. In doing this, you will have to put your ear right up to the speaker and listen quite carefully.

4-7 Review Questions

1. How do you tell the difference between a sound if trouble and an audio power-amplifier trouble?
2. In a solid-state set, the collector supply voltage is 24 volts. What would you expect the collector-to-emitter voltage of the sound if stage to be?
3. In Question 2, what would you expect the emitter-to-base voltage to be?
4. The circuit diagram of the set in question shows the collector supply voltage to be 28 volts and the collector-to-emitter voltage to be 24 volts. If you measured a collector-to-emitter voltage of 21 volts, would you think the stage was working normally?
5. The collector-to-emitter voltage in Fig. 4-9 measures zero, and the power supply voltage is 18 volts. What would you think was wrong?
6. A certain set has no sound but the picture is all right. What would you do to try to pin down the cause?
7. In the set in Question 6, the background noise varies with the volume control. Which stages would you suspect?
8. When you check the set in Question 7, you find that the audio circuits are working and the voltages in the sound if amplifier are normal. What would you do next?

9. How would you try to check a sound detector like the one in Fig. 4-9?
10. What would you do if the set had no sound at all and all of the stages had normal voltages? (The picture is fine.)
11. How would you check the circuit shown in Fig. 4-8?
12. In a vacuum-tube sound if amplifier, the screen-grid voltage is 150 volts and the plate voltage is zero. What would you suspect has happened? (Use the circuit in Fig. 4-4.)
13. How would you check the tuned transformers in a sound if circuit?

CHAPTER **5**

Troubleshooting Video Failures

If a television set has a component failure that affects *only* the video reproduction, then the trouble must lie in the video amplifiers or in the picture tube. The usual symptoms of such a video trouble will be:

1. Sound normal
2. Raster normal
3. Video absent, weak, or distorted

When these symptoms are observed, the shaded stages in Fig. 5-1 should be suspected.

5-1 Television Picture Tubes

The cathode-ray picture tube is perhaps the key component in a television receiver. Certainly tv as it is known today would not be possible without such a device.

Fig. 5-2 illustrates how a cathode-ray tube works. A small voltage is applied to the *heater* (1), which is somewhat like the wire in an electric toaster. When voltage is applied, the heater gets red hot. The heater of a tube can be seen glowing near the base.

When the heater gets hot, it in turn heats up the *cathode* (2). The cathode element is coated with a substance that will *emit* (give off) electrons when it is heated. This action can be imagined by thinking of a pan of water on an electric stove. As the stove element (heater) gets hot, it warms the pan (cathode frame) and the water (cathode coating), which in turn releases (emits) steam (electrons).

In normal operation, the *first anode* (3) has several hundred volts connected to it, and the *second anode* (4) has many thousands of volts connected to it. These high voltages attract the electrons toward the *phosphor coating* (5) on the screen (6) about like a large exhaust fan would draw the steam from the pan of boiling water.

When the electron beam strikes the phosphor coating, the phosphor glows, creating a small spot of light on the screen.

Back in the neck of the tube is one component that has not yet been considered, the *control grid* (7). This element normally has a negative voltage applied to it. The negative voltage neutralizes the effect of part of the positive first-anode voltage and therefore is able to *control* the number of electrons in the beam. If fewer electrons strike the phosphor coating, then less light is produced. In this way, the control grid can vary the brightness of the spot.

In actual practice, the video voltage can be used to vary the spot brightness on the screen by applying it to the control grid and holding the cathode voltage fixed. It works just as well to hold the control-grid voltage fixed and vary the spot brightness by connecting the video voltage to the cathode. You may find either method used in a modern television set.

By controlling the electron beam current, the spot brightness may be varied. Since all of the beam current comes from the cathode, one might expect the *brightness control* to be located in the cathode circuit. This is usually the case, but there are other ways to adjust brightness that will be examined later.

Focusing of the picture can be done by adjusting the voltage on one of the picture-tube anodes. When this is done, the focus control adjusts the size of the spot on the screen.

All of the picture-tube elements that are located in the neck of the tube are used to form and control the electron beam. These parts are

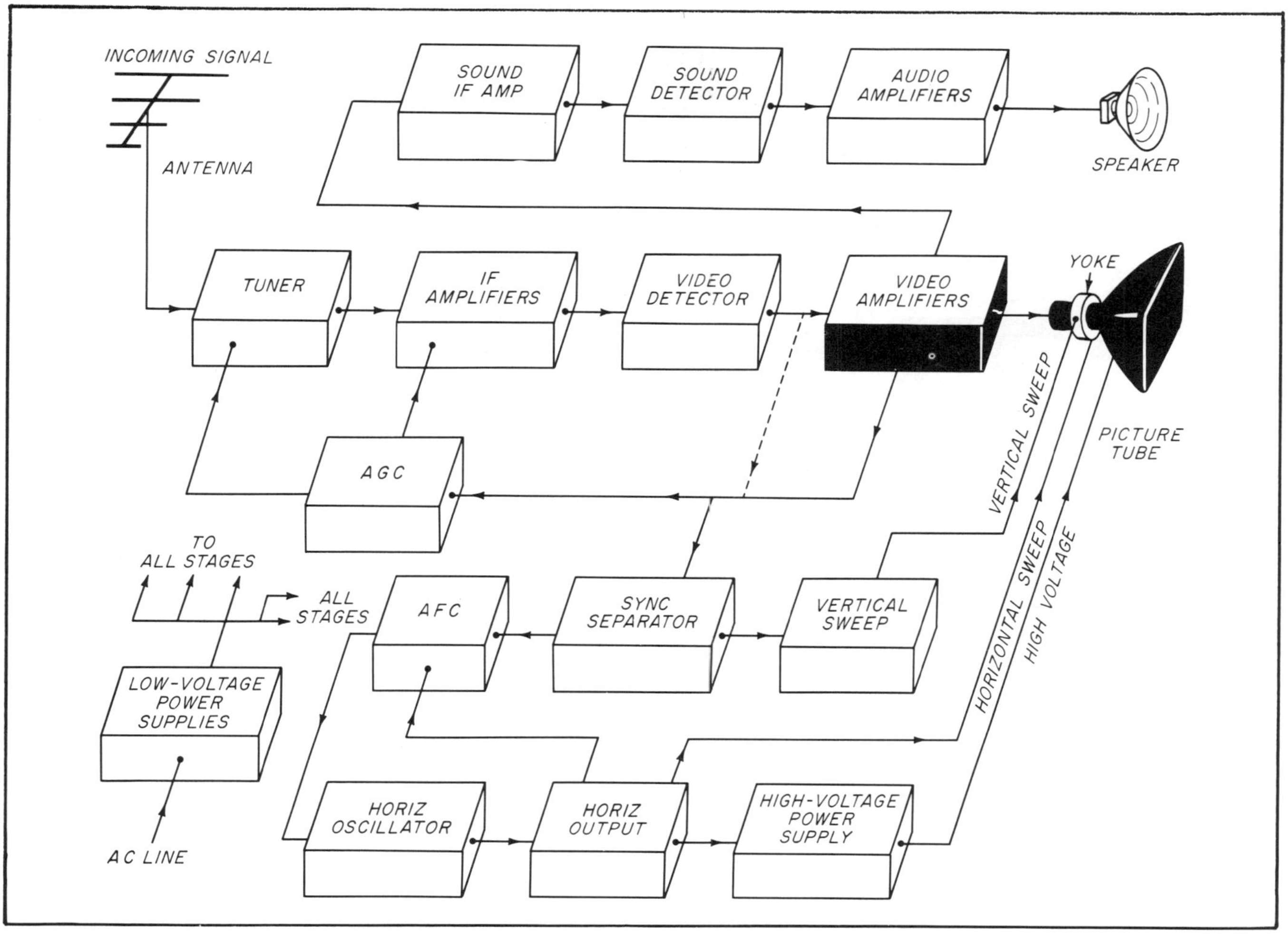

Fig. 5-1. Block diagram of typical black-and-white receiver.

frequently considered as a group and called the *electron-gun* assembly.

It is perhaps worth mentioning at this point that the *deflection yoke* is mounted on the neck of the picture tube. The coils in the yoke function as electromagnets and position the spot in the desired location throughout the scanning process.

The construction of color picture tubes is not much different from that of black-and-white picture tubes. Fig. 5-3 shows pictorial sketches of two color tubes.

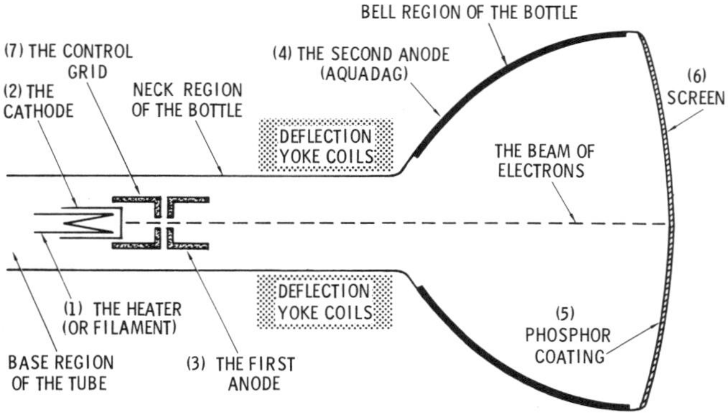

Fig. 5-2. A black-and-white picture tube.

The color tube in Fig. 5-3A has three separate electron guns, one for each of the colors red, green, and blue. Also, the phosphor is different from that in a black-and-white tube. In this color tube, the phosphor is made up of about 350,000 three-color groups called *triads.* Each electron gun is aimed so that it will activate only its own color dots. Just behind the phosphor of the color tube is a screen called a *shadow mask.* This mask has one hole for each color triad. It ensures that the beams will activate only one triad at a time. If a color tube is observed very closely when the set is turned off, the tiny color dots can be seen.

A second type of color picture tube is growing in popularity. It is the *in-line* tube (Fig. 5-3B). It uses three cathodes and a grid system that forms three side-by-side electron beams. The shadow mask has slots rather than round holes, and the

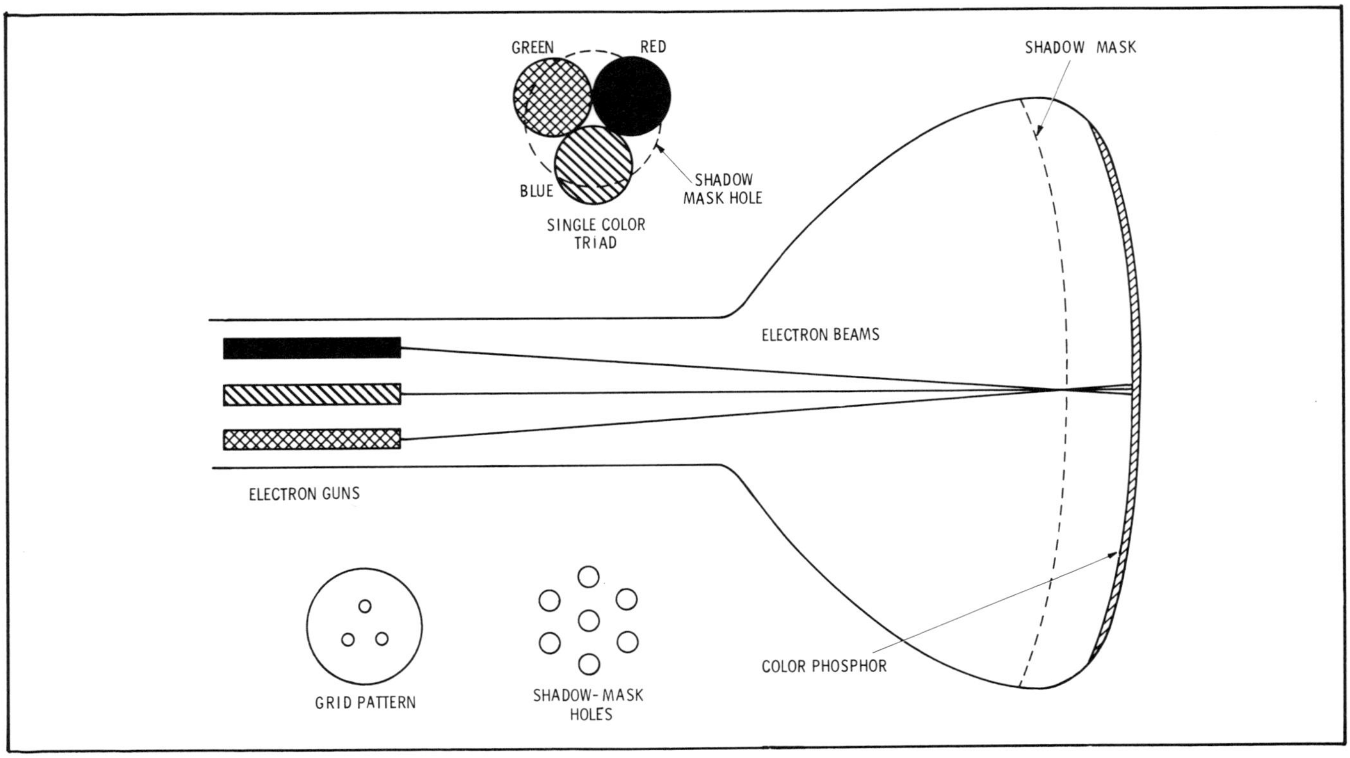

(A) Delta system.

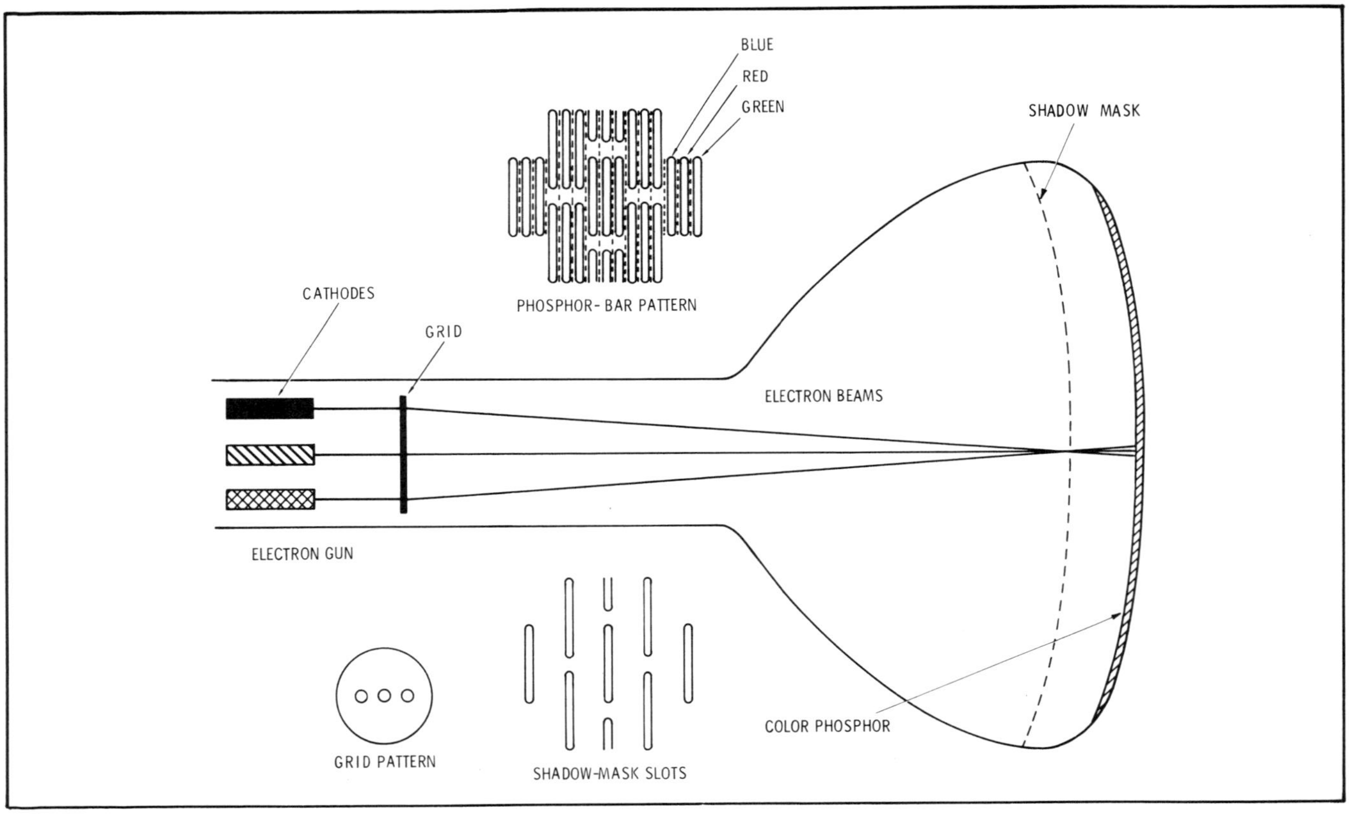

(B) In-line system.

Fig. 5-3. Construction of color picture tubes.

phosphor is put on in a bar pattern. The overall result is a tube that is bright, has high resolution, and is easy to adjust properly. Some in-line tubes come with the yoke and purity magnets permanently mounted and preadjusted.

Another type of tube worth mentioning is the *Trinitron* picture tube used in some Japanese-made receivers. This color tube uses only one gun to produce the three beams. Extensive information on the *Trinitron* tube can be found in *Color TV Training Manual* published by Howard W. Sams & Co., Inc.

5-2 Video Amplifiers

A television receiver may have only one video-amplifier stage, or it may have several. Vacuum-tube and hybrid sets tend to have fewer video-amplifier stages than do solid-state sets. In any case, the first video-amplifier stage performs several functions. These functions are:

1. Video amplification
2. Takeoff point for the sound if signal
3. Takeoff point for the sync signals
4. Takeoff point for the agc signal
5. In many sets, the chroma (color) signal is also taken off here.

If this stage fails, it will normally cause symptoms to appear in the sound, video, and sync outputs. Consequently, if the sound and raster are normal, then the first video amplifier can be assumed to be working properly.

Fig. 5-4 shows a simplified three-stage video amplifier. The first stage is an emitter follower, and the sound, sync, and agc are taken off before

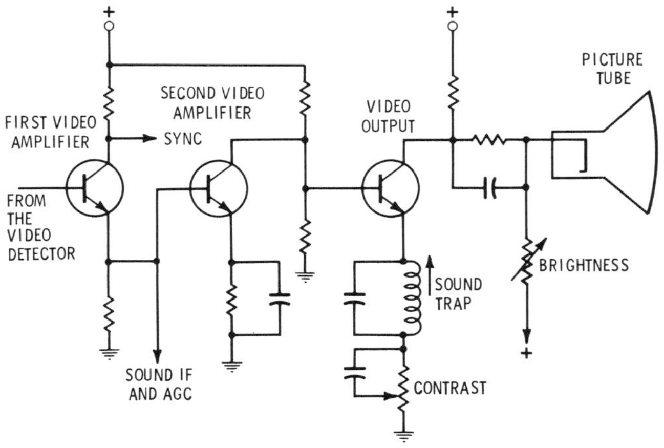

Fig. 5-4. A simplified video-amplifier section.

the second stage. A failure in the second video amplifier or video-output stage would cause a loss of video but would probably not affect the sound or raster.

A normal troubleshooting procedure would be to measure the voltages and compare them to the normal values. Also, the video signal could be observed with an oscilloscope at the input and output of each stage. When the earliest stage that does not pass the signal is identified, the individual components can be checked to isolate the problem. As in other circuits, the active device (tube or transistor) is the thing which most often fails, and it should therefore be checked first.

It is fairly common practice to use direct coupling in video amplifiers as is shown in Fig. 5-4. Notice that not only is the video signal directly coupled but the bias on each stage is directly affected by the preceding stages. For example, the base bias voltage on the output amplifier is the same as the collector voltage on the second stage. This type of coupling often causes some confusion in troubleshooting.

To illustrate this point, imagine that the second video amplifier fails due to an emitter-to-collector short in the transistor. This failure would cause the collector voltage at the second video stage to drop to a value much below normal. Since this collector is tied directly to the base of the output stage, the output-amplifier base voltage would drop also. When the base voltage dropped, the base current would drop, and the output transistor would cut off. This would cause the collector voltage of the output stage to increase to near the value of the collector supply. This increase in output collector voltage would cause a decrease in picture brightness, so the external symptoms would be normal sound, no video, and a dim raster.

This would be recognized as a video-amplifier failure, and the circuit should be checked with a volt-ohmmeter. When the output-stage collector voltage is measured, it will be too high, and if no further check is made, a lot of time may be wasted looking for trouble in the output stage.

To avoid this kind of confusion, it is sensible to check each stage of a direct-coupled amplifier before jumping to conclusions. If this is done in the case above, it will be found that the collector voltage of the output stage is high, the collector voltage of the second video amplifier is low, and the collector voltage of the first video amplifier is normal. In almost all cases, it is *the earliest abnormal direct-coupled stage* that causes the trouble.

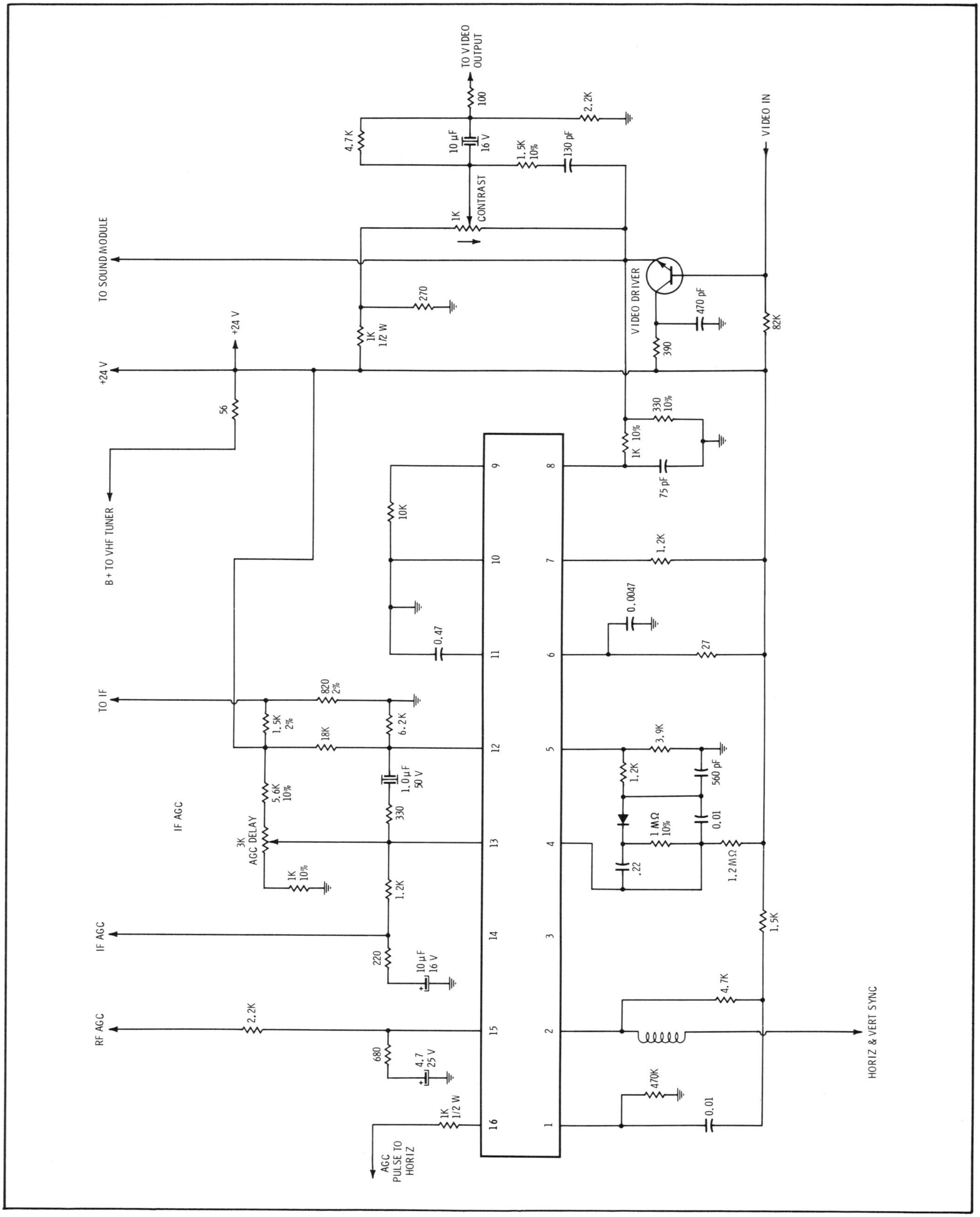

Fig. 5-5. An integrated-circuit video processor.

The sound-trap circuit in the video-output stage should also be examined. The circuit is tuned to the 4.5-MHz sound if frequency. Its purpose is to prevent the sound signal from being seen in the picture. If there is sound present in the picture, it may indicate that the trap is misaligned or that the capacitor in the trap is shorted. If the capacitor is open-circuited, the video gain will be reduced at higher frequencies, which will result in a serious loss of picture detail.

The circuit shown in Fig. 5-4 is an extremely simplified version of a video amplifier. To get a more practical familiarity with typical circuits, video amplifiers in the set diagrams at the back of the book should be examined.

In many recently produced sets, an integrated circuit is used to do video processing. This is a trend that will undoubtedly continue to spread. Fig. 5-5 shows the schematic diagram of one such arrangement. The video processor itself provides all of the signals required for other stages in the set.

Some sets employ *video peaking* to control the amount of fine detail that is presented on the screen. The amount of video peaking may be fixed at manufacture, or it may be adjustable by a front or rear panel control that is usually called *peaking* or *sharpness.*

A failure in the picture tube can result in a loss of video. When the picture tube is suspected, it can be checked with a crt (cathode-ray tube) tester or by substituting a known good tube.

5-3 Contrast Control

Varying the video voltage applied to the picture tube causes the brightness of the spot to vary. When the spot is completely cut off, the screen is dark. When the spot is fully on, the screen is producing maximum light. The difference between the light and dark parts of the screen is what is called *contrast.* Since the variation from light to dark depends on the variation in video voltage, it can be seen that the greater the video variation is, the greater the picture contrast will be.

The variation in video voltage depends very much on the gain of the video amplifiers. Therefore, the contrast control in the set is actually a video-gain control and will always be somewhere in the video-amplifier circuitry. In most cases (but not all), the contrast control is found in the video-output stage.

5-4 Summary of Video Troubles

As previously discussed, when the sound and raster of a set are normal while the video is missing, the trouble must lie in the picture tube or in the video amplifier. However, since the sound, sync, and agc are taken off at the first video stage, a failure in this stage will usually also affect the sound output.

5-5 Review Questions

1. Explain why the video amplifiers and the picture tube are the only parts of Fig. 5-1 that are shaded.
2. What does crt mean?
3. What is the function of the heater in the picture tube?
4. Explain in your own words how a crt works.
5. Use the technical data for actual sets and determine typical values of second-anode voltage for black-and-white and color picture tubes.
6. How many holes are there in a typical shadow mask?
7. What does the brightness control of a set do in the circuit? What does the contrast control do?
8. What is the difference between picture brightness and contrast?
9. What is an electron gun?
10. How many cathodes does a color picture tube have?
11. Would you expect a tube-type set or a solid-state set to have more stages of video amplification?
12. What are the functions of the first video stage?
13. In which stage of a set would you expect to find the contrast control?
14. Suppose that the second video stage in Fig. 5-4 had a burned out transistor (the collector-emitter junction is open circuited). What would be the external symptoms?
15. In Question 14, would the collector-emitter voltage of the second video stage be high or low?
16. In Question 15 would the collector-emitter voltage of the video-output stage be low, normal, or high? Explain your answer.
17. The picture tube in Fig. 5-4 has no video, but the sound and raster are normal. Using

an oscilloscope, we discover that the video is normal at the base of the video output stage but not at the collector. What component failures could cause this?

18. In Question 17, suppose that the collector-emitter voltage of the output stage is zero. What component failures could cause this? How would you check them?
19. What would be the circuit indications if the emitter resistor in the second video stage (Fig. 5-4) was open?
20. The picture tube in Fig. 5-4 checks OK, and an oscilloscope reveals that the video at the collector of the output amplifier is normal, but there is no video on the screen. The sound and raster are normal. What do you think the problem is? How would you verify your opinion?

CHAPTER 6

Picture IF Problems

The block diagram in Fig. 6-1 shows that both the video and sound pass through the picture if amplifiers. Therefore, a failure in this section would be expected to affect both the picture and the sound. This is indeed what happens.

The picture if section usually contains either two or three if amplifier stages. Fig. 6-2 shows a three-stage picture if block. The video detector is also shown in this illustration. It is often difficult to distinguish between failures in the detector

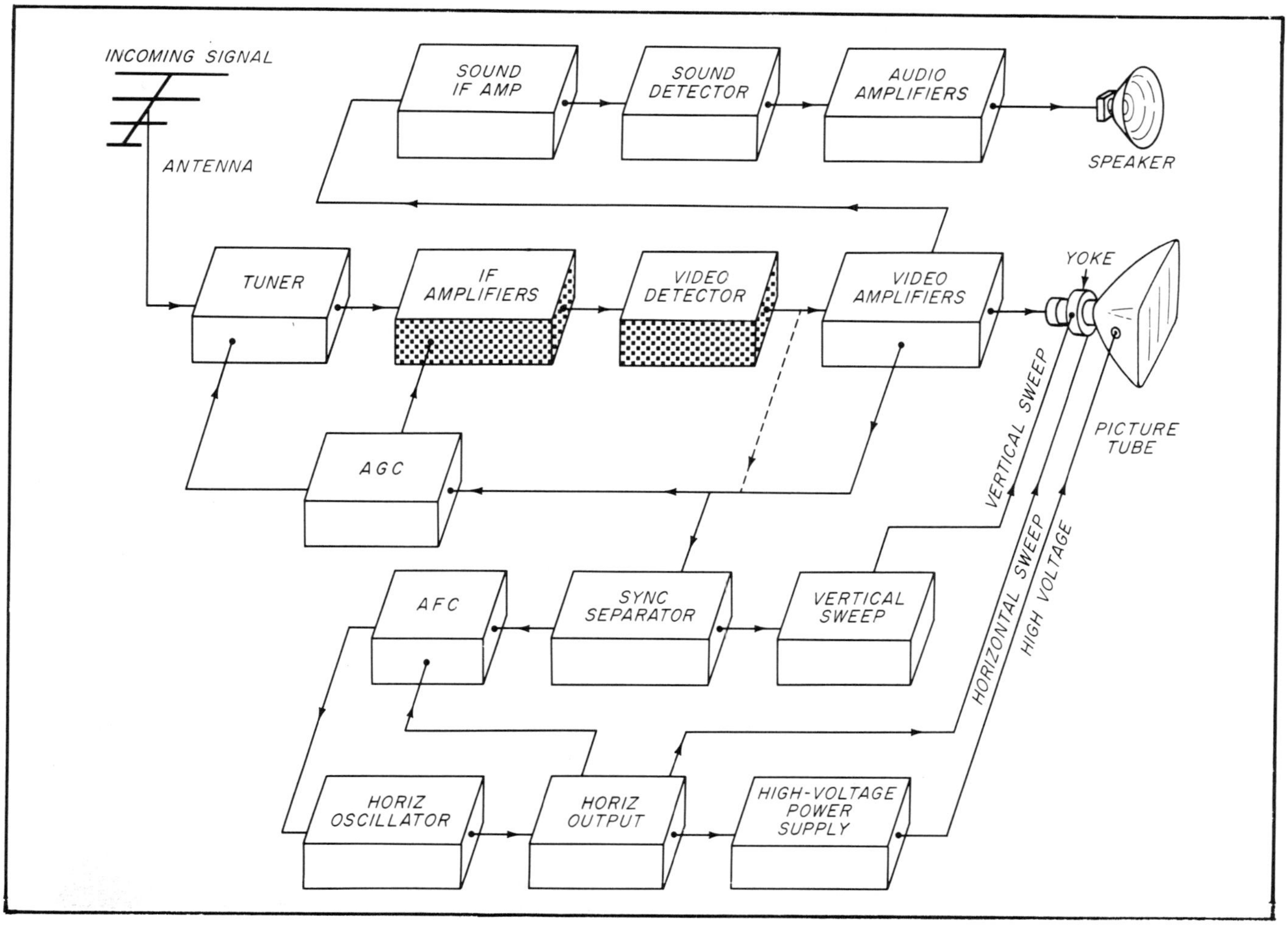

Fig. 6-1. Block diagram of a typical black-and-white receiver.

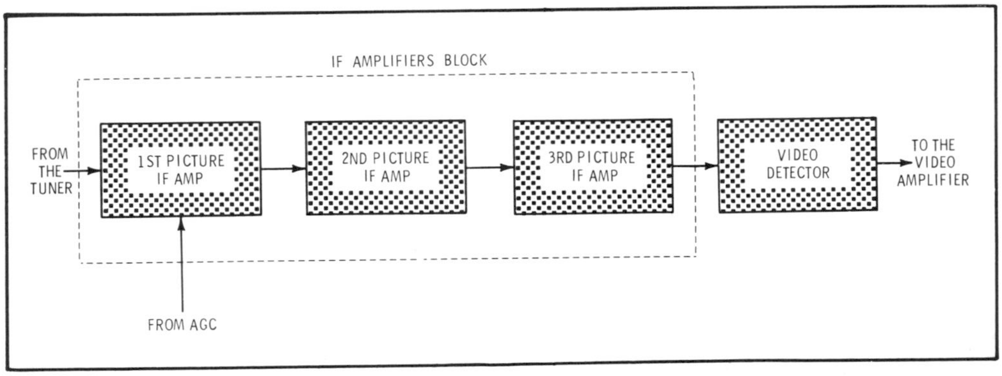

Fig. 6-2. The picture if section.

and failures in the third if stage. For this reason, consider the detector to be a part of the picture if amplifier section.

6-1 Recognizing IF Amplifier Trouble

Check the blocks in Fig. 6-2 one at a time, and consider what happens if they fail. If the video detector fails, both the sound and the video will be dead. By listening very carefully to the sound while varying the volume control, you will hear:

1. Some soft background hum
2. Some soft background noise (usually a hissing sound) that goes up and down with the volume control
3. No program sound

The picture will have a normal-appearing scanning raster and no video. If the contrast control is varied, there will be little noticeable change in the picture.

The third picture if amplifier will produce the same symptoms if it fails. There may be very slightly more hissing in the sound, and at full contrast just a hint of snow may appear in the picture. Fig. 6-3 shows the screen of a set that has a failure in the video detector or in the third if stage.

Fig. 6-3. No video, normal raster.

Skip over the second picture if stage for a moment, and consider a failure in the first if amplifier. The first picture if amplifier stage gets its signal from the tuner and a control voltage from the agc circuit. The agc operation will be examined in some detail in Chapter 8. For the moment ignore it.

If the first picture if stage fails, both the sound and the picture will be affected. However, the effect will be quite different from the effect caused by failure of the third if amplifier. Even if the first if stage fails completely, some picture or sound will usually leak through. Also, since there is some noise generated inside each amplifier stage, some snow can usually be seen when the contrast is turned up.

The symptoms usually associated with a defective first picture if stage are:

1. Very weak video with some snow
2. Very weak sound with a hissing background noise
3. Both 1 and 2 at the same time

Fig. 6-4 shows the symptoms that should be seen if the first picture if stage fails. Notice the weak, washed-out looking video and the snow.

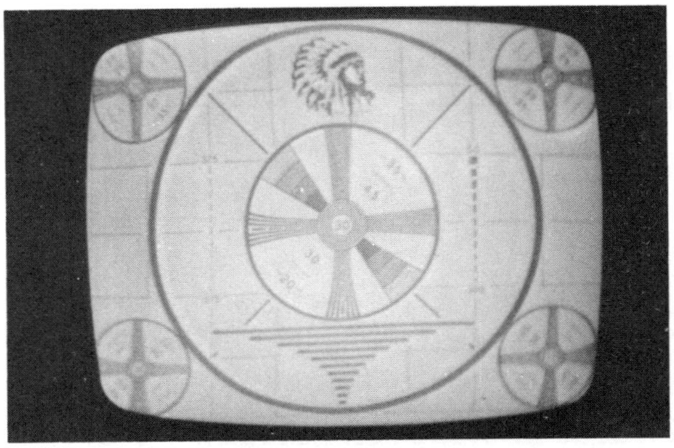

Fig. 6-4. Weak video caused by failure of if amplifier.

In some sets, there are only two if amplifier stages. In that case, there is no problem with a middle if stage. In other sets, there are three if stages, so problems can arise in the middle one.

Because the second if stage is between the first and third ones, its failure symptoms can be *similar* to those of either of the others. For example, there may be some picture or sound leaking through—but then again there may not. Also, there may or may not be some snow in the picture. And, there may or may not be some hissing noise in the sound.

Perhaps the best way to go is to look at the symptoms and decide if you think the trouble is in the first part of the if section (weak video or sound, some snow and hiss), or in the last part (no video or sound, little snow and hiss). If the first part is suspected, check both the first and second stages. If the last part is suspected, check the second and third stages.

A good way to isolate trouble to the if section is to:

1. Vary the contrast control while watching the picture very closely. Any noticeable change in the picture means that the video amplifiers are probably OK.
2. Flip the channel selector from station to station. If the if stages are OK, some change should be seen in the picture (provided the video is working all right).

In many sets, particularly solid-state ones, if the channel selector is set *between* channels, you will see snow on the screen if the if section is working.

6-2 Video-Detector Circuitry

The video detector performs two services in a television receiver. It converts the video from the picture if signal back to actual video, and it also converts the sound signal to the sound if frequency. Fig. 6-5 shows a simplified video-detector circuit. The input signal from the picture if stages is usually detected by a solid-state diode.

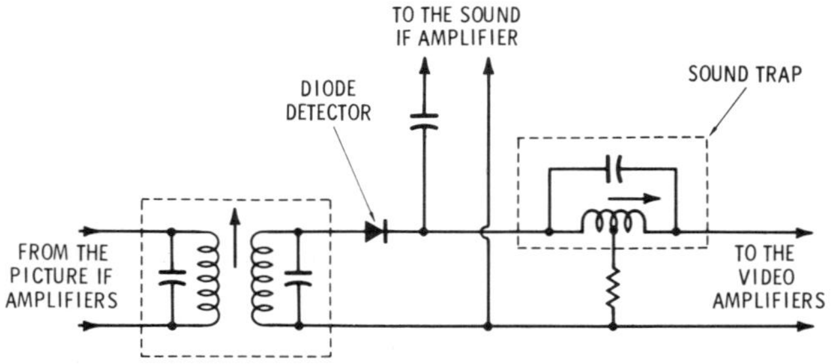

Fig. 6-5. Simplified video-detector circuit.

In some sets, the sound if signal is separated from the video signal at the video detector. This is the case in Fig. 6-5. When the sound is removed at this point, there will be a sound trap like the one shown. This trap prevents the sound if signal from entering the video amplifiers.

In other sets, the sound if signal is taken off *after* the first video stage. In such a case, the sound trap would not be in the detector circuit at all. It would be just after the sound takeoff point between the first and second video amplifiers.

A few sets use separate sound and video detector circuits. Both detectors are about like the one shown in Fig. 6-5.

In any case, the video detector will be a very simple circuit. Because of its simplicity, it fails only rarely. It should not, however, be completely forgotten.

The tuning of the coils and traps in the detector circuit is critical and difficult. These adjustments should be changed only by an expert technician who is very familiar with the procedure.

6-3 Picture IF Amplifier Circuits

The picture if amplifiers must amplify a band (or range) of signal frequencies from about 41 MHz up to about 46 MHz. The sound and color signals are near the lower end of this band (41.25 and 42.17 MHz, respectively), and the video is near the upper end (45.75 MHz). As a result, the various picture if stages have rather complex tuning arrangements. Fig. 6-6 shows a typical if amplifier section. Notice that the video-detector circuit in this diagram is considerably more complicated than the simplified one shown in Fig. 6-5.

As in the case of the detector, the tuning adjustment of an if amplifier is critical and difficult. These adjustments should not be changed except by expert technicians who have specialized instruments available.

Vacuum-tube if amplifier circuitry is found in some sets. Fig. 6-7 shows one of the popular vacuum-tube circuit arrangements.

As the move to more and more large-scale integration has proceeded, manufacturers have included the if circuits as well as others. Fig. 6-8 shows the module diagram of a set with an integrated-circuit if section and an integrated-circuit sound section.

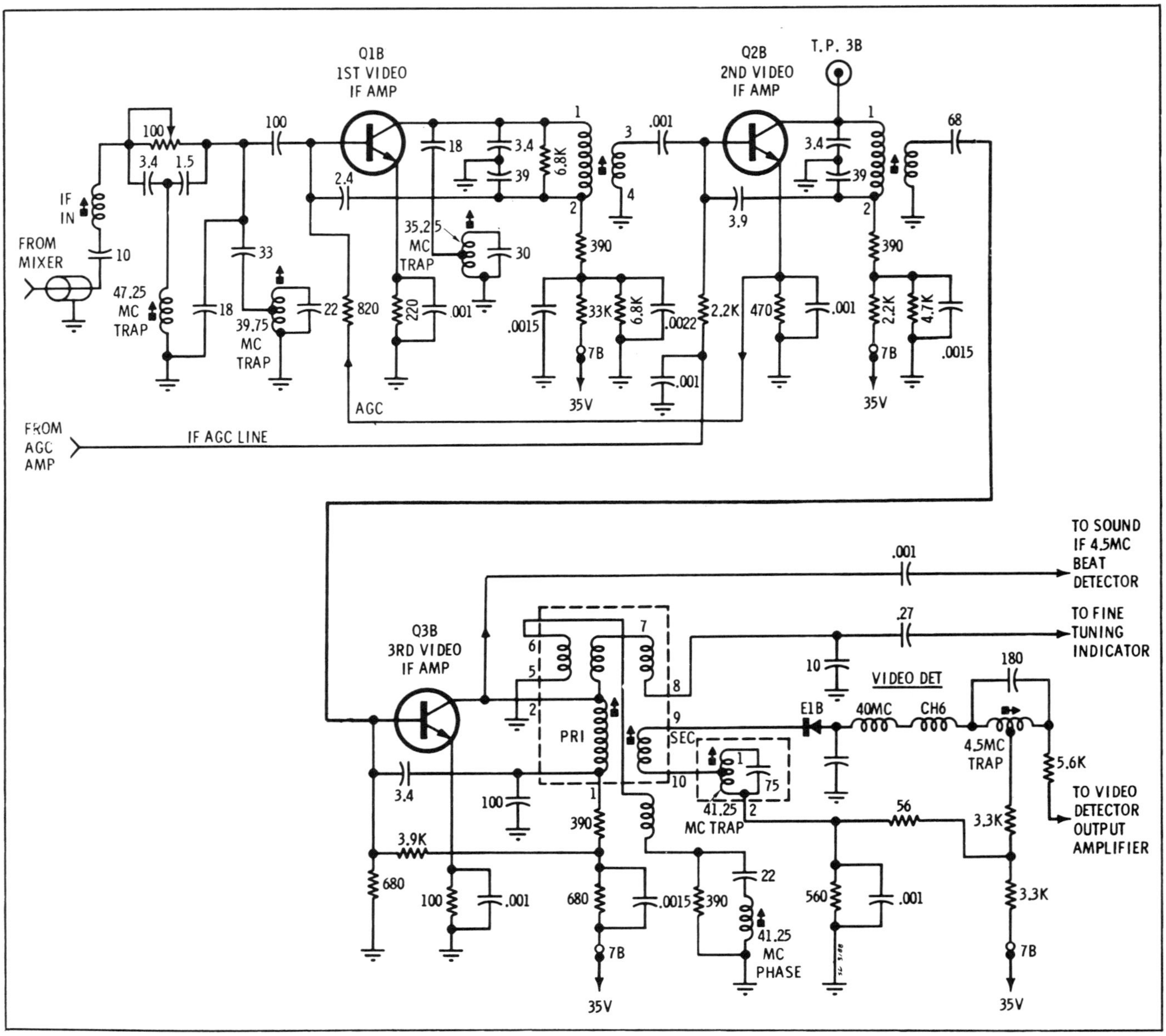

Fig. 6-6. Solid-state picture if circuit.

6-4 Troubleshooting the IF Circuitry

Once a problem has been diagnosed as an if-section failure, troubleshooting can be carried out in the same way as with any other tuned amplifier. To illustrate the procedure, a couple of examples follow:

Suppose you have a set in which the raster appears normal but there is no video, no sound, no snow, and only the faintest hissing noise with the volume turned all the way up. Varying the contrast control does not seem to change anything. Neither does flipping the channel selector from station to station. This problem would probably be diagnosed as being in the late if amplifiers or in the video detector.

Suppose that Fig. 6-6 is the if section of the set. The voltmeter should be used to start checking circuit voltages:

1. Measure the power-supply voltage at test point 7B—it reads about 35 volts.
2. Measure the collector-to-emitter voltage of Q3B (the third if amp)—it reads zero.
3. Measure the emitter-to-ground voltage of Q3B—it reads about 3 volts.

Based on these readings, it could be concluded that the power-supply voltage is OK and that there is

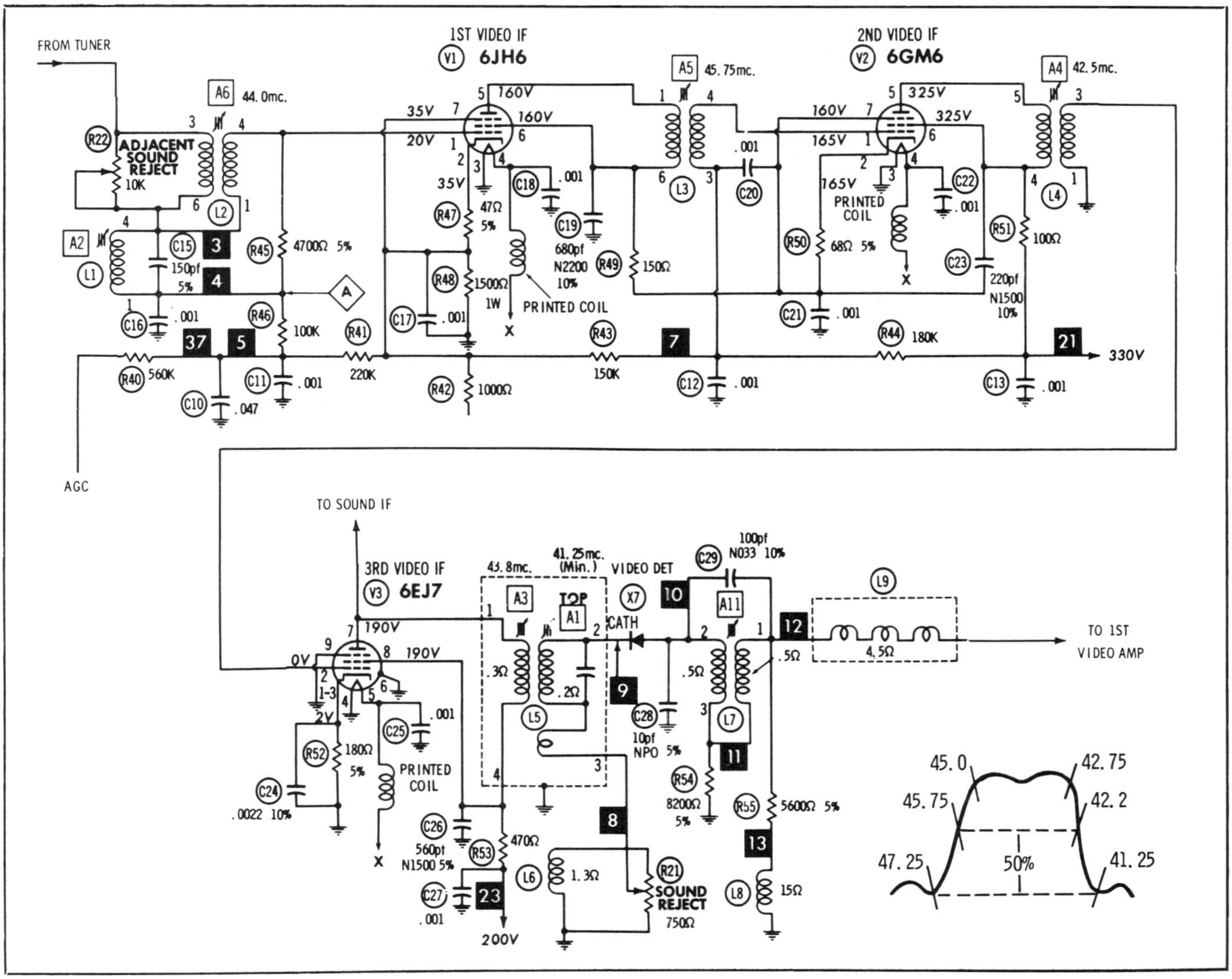

Fig. 6-7. Vacuum-tube picture if circuit.

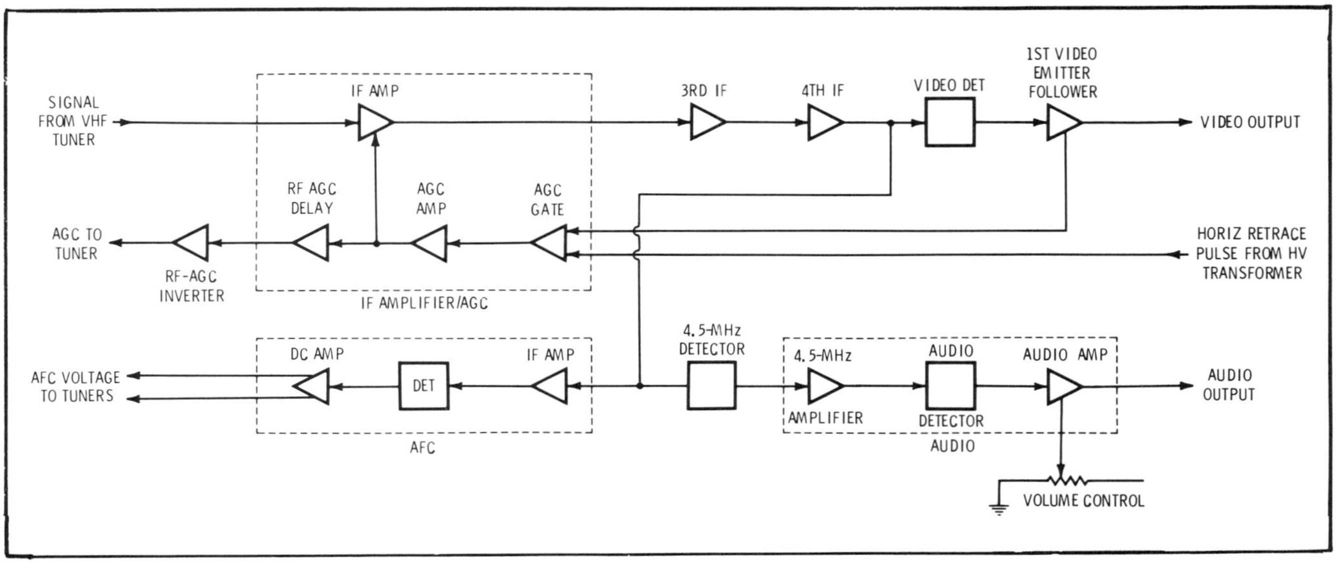

Fig. 6-8. Partial diagram of a set with integrated-circuit if and sound sections.

some emitter-circuit current. (If there were not, the emitter-to-ground voltage would be zero.) But there is no voltage from collector to emitter. Hmmmmm—acts like a collector-to-emitter short in Q3B—better check it. After the transistor is removed, it is indeed found to be shorted. Replacing Q3B with a new one restores normal set operation.

Now look at another set. The symptoms are the same except that there is a little bit of program sound (very weak), and a very faint picture can be seen when the contrast is all the way up.

These symptoms sound like an early if problem. Fig. 6-7 is the if section of this set. Before pulling the chassis out, check the 6JH6 first if tube. Replacing it with a new one does not seem to help any, so put the original tube back. Try the 6GM6 second if tube next. Well, that is what the trouble was all the time. If the tube replacements had not cured the trouble, it would have been necessary to take out the chassis and measure the circuit voltages as before.

6-5 Summary of IF Trouble Symptoms

Problems in the if section will normally affect both picture and sound. The raster will appear to be normal. The picture and sound may be completely dead or just very weak. There may be some snow and noise present, but they will not be severe.

6-6 Review Questions

1. Describe the symptoms you would expect if the video detector in a receiver failed.
2. What are the symptoms that you would expect from a failure of the first picture if?
3. If Q2B in Fig. 6-6 had an open emitter, what would you expect the collector-to-emitter voltage to be?
4. If the third picture if amplifier in a tv set fails, what symptoms would you expect?
5. Diode X7 in Fig. 6-7 is burned out. What symptoms would you expect?
6. The collector-to-emitter voltage of Q1B in Fig. 6-6 is about 35 volts. What do you think is wrong?
7. What would be the picture and sound symptoms in Question 6?
8. Resistor R51 in Fig. 6-7 is an open circuit. What value of plate-to-cathode voltage would you measure at V2?
9. What would be the picture and sound symptoms in Question 8?
10. A tv set has no picture, no sound, and no snow. Which stages would you suspect of having failed?
11. Which stage in Question 10 would you check first?
12. In the if-section diagram in Fig. 6-6, the collector-to-emitter voltage of Q3B is zero. List each component that could cause this symptom and the nature of the failure. For example: Q3B—shorted collector to emitter.
13. The plate-to-cathode voltage of V1 in Fig. 6-7 is zero. List the components that could cause this problem and the nature of the trouble.
14. Video detector E1B in Fig. 6-6 is known to be burned out. There is no video on the screen but the sound seems normal. How do you explain this?
15. The sound in the set of which Fig. 6-6 is a part is dead. The picture is normal, but the fault is known to be in the third video if amplifier. Which component in Fig. 6-6 is at fault?
16. A vacuum-tube tv set has the if section shown in Fig. 6-7. If both the sound and picture are dead, could the trouble be in the video-detector stage? Why?
17. A set which has an if section like Fig. 6-6 has a strange symptom. There are wavy lines in the picture that seem to wiggle in step with the program sound. In which stage do you think the problem lies?

CHAPTER 7

Tuner Troubles

The tuner in a television receiver is a little like the quarterback of a football team. The quarterback selects the play that the team will use, and the tuner selects the channel that the set will work on. In doing this, the television tuner has three basic functions:

1. It selects the desired channel.
2. It adjusts the level of the signal.
3. It converts the signal to the intermediate frequency (if).

These basic functions should be kept in mind as

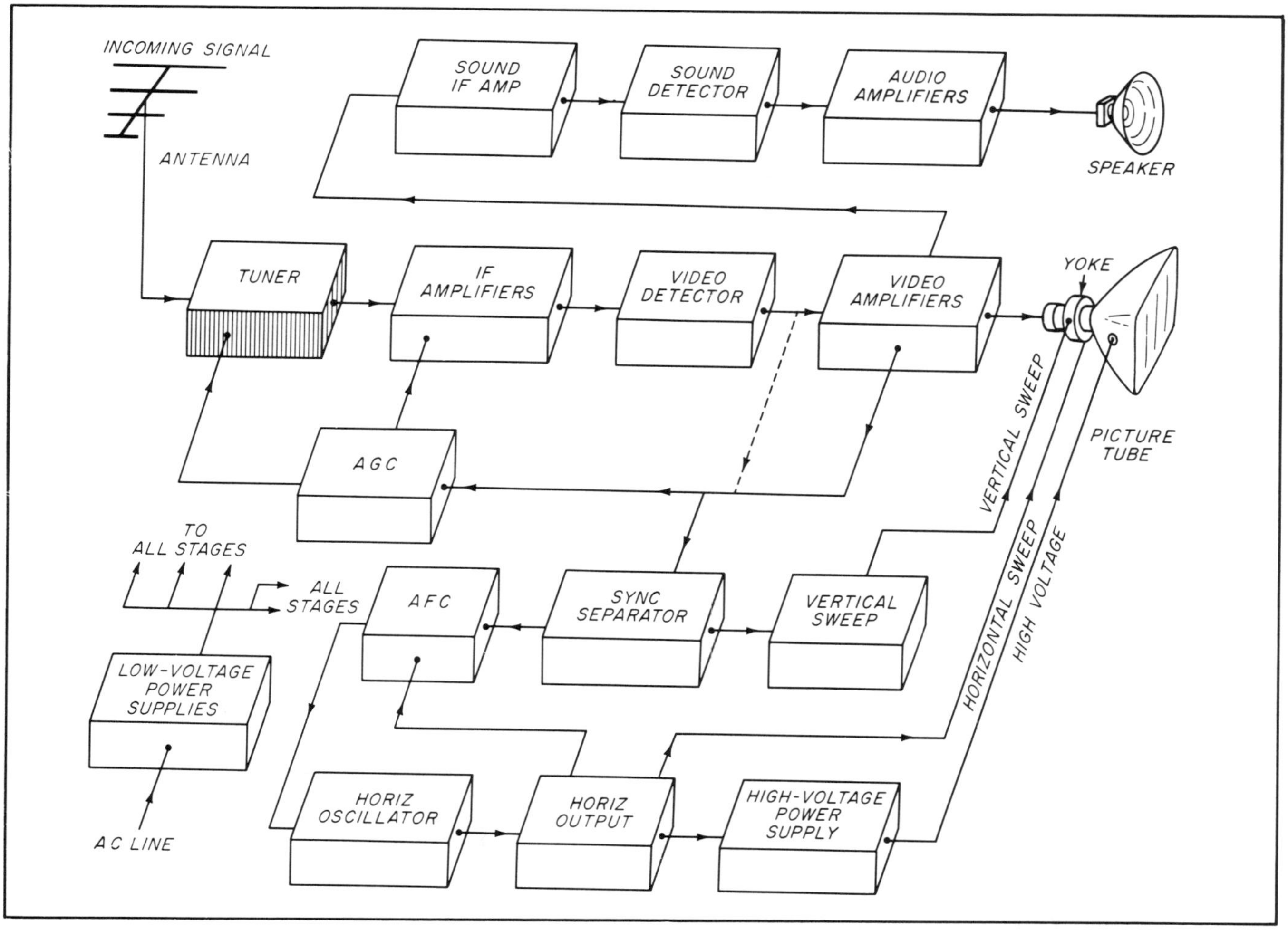

Fig. 7-1. Block diagram of a typical black-and-white receiver.

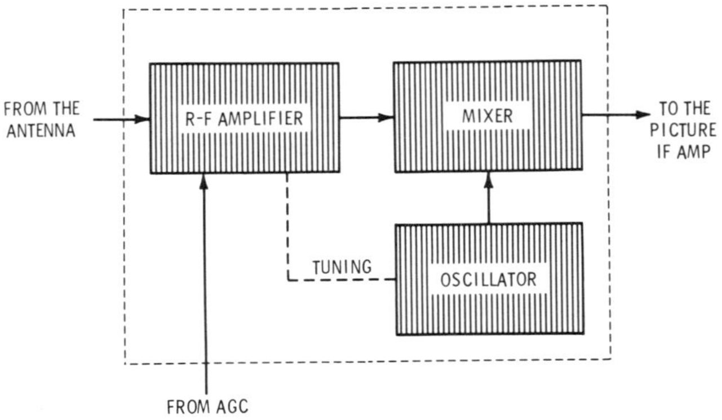

Fig. 7-2. The tuner stages.

we discuss the tuner's operation. Fig. 7-1 shows the block diagram of a typical black-and-white tv receiver. Notice that the tuner is located between the antenna and the picture if amplifiers. Inside the tuner block, there are normally three stages. These stages are:

1. The rf amplifier
2. The oscillator
3. The mixer

Fig. 7-2 shows how the tuner stages are connected to each other.

7-1 Recognizing the Symptoms

If the mixer stage fails, the symptoms will look very much like those caused by a failure in the first if amplifier. There will be some snow on the screen, and there may or may not be some very weak video and sound.

In many sets (solid-state ones in particular),

Fig. 7-3. Mixer failure symptoms.

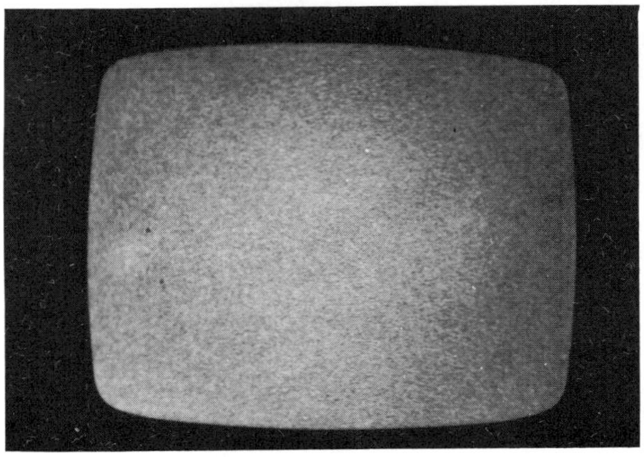

Fig. 7-4. Rf-amplifier failure symptoms.

you can check the if amplifiers by switching the channel selector to a point between two channels. If some snow appears on the screen, the if amplifiers are all right. However, this test will not work with all sets. Fig. 7-3 shows the screen of a set with a bad mixer transistor.

The rf-amplifier stage selects the desired channel and adjusts its level. If this stage fails, the symptoms will be about the same as those described for a bad mixer, except that the snow will be much more severe. Fig. 7-4 shows the screen of a set with a defective rf amplifier. It should be kept in mind that there may be some weak video and sound present. The snow associated with an rf-amplifier failure is severe enough that it should be easy to recognize.

The television antenna is really part of the rf-amplifier circuit. If there is a problem in the antenna, the symptoms will look like those of a bad rf amplifier. So the antenna connections should be checked when the heavy snow associated with a bad rf amplifier is seen.

The oscillator signal is mixed with the rf-amplifier signal to produce the picture if signal. If the oscillator fails, there will be no if signal; and since most of the snow seen when the rf amplifier fails comes from the oscillator, there will normally be

Fig. 7-5. Oscillator failure symptoms.

very little snow. Fig. 7-5 shows the screen of a set in which the oscillator has failed.

In summary, a failure in the tuner will give these symptoms:

1. If the mixer fails, there will be some snow, and there may or may not be some weak video and sound. The symptoms look a lot like those of first-if trouble.
2. If the oscillator fails, there will be little snow and no video or sound. The symptoms look a lot like those of third-if trouble.
3. If the rf amplifier fails, there will be severe snow, and there may or may not be some weak video and sound. *Do not forget to check the antenna and its connections.*

7-2 Tuning a TV Receiver

Before looking at tuner circuitry, consider how a set is tuned to various channels. If every stage from the rf amplifier through all the if stages to the video detector had to be tuned each time the channel was changed, it would be awfully complicated. You have probably seen pictures of old-time radios with lots of tuning knobs. Those old radios were called tuned-radio-frequency (trf) sets, and you did have to tune every rf stage separately.

A television set uses a tuning method called *heterodyning*. With this method, all of the stages except two are permanently tuned to an *intermediate frequency* (if). In a tv set, the intermediate frequency is about where channel 1 would be if there were such a channel. So, you can think of all parts of the set except the tuner as working on channel 1. The tuner itself works on all of the channels except channel 1.

To understand better how this heterodyning system works, let's go through an example. Suppose that it is desired to watch channel 6, which is on a frequency of about 83 MHz. The rf amplifier is tuned to this frequency, and it picks up channel 6. The rf amplifier amplifies the signal and feeds it to the mixer stage.

At the same time that the rf amplifier is tuned to 83 MHz, the oscillator is tuned to about 129 MHz. The oscillator generates a signal of its own at this frequency and feeds it to the mixer. Inside the mixer, the two separate signals are *heterodyned*—that is, they are mixed together in a way that creates a signal that is equal to the *difference* between them. This difference in signal is the intermediate frequency (about 46 MHz).

If we switch to a new channel, both the rf amplifier tuning and the oscillator tuning are changed. But, the difference between them stays at the intermediate frequency. In this way, only two stages at a time must be tuned instead of six or more as would be the case with a trf circuit.

You can probably guess that it is important for the frequency of each channel to be kept very accurately fixed. Also, the tuning circuits of the rf amplifier, oscillator, mixer, if amplifiers, and detector must not be tampered with, especially by the beginning technician, if the system is working well.

The adjustments necessary to get the whole set working right are very delicate and complicated. Only a thoroughly trained, specially equipped technician should undertake to align or adjust a television receiver.

7-3 RF Amplifier Circuits

Radio-frequency amplifiers come in a fairly wide variety of circuits. They may employ ordinary bipolar transistors as shown in Fig. 7-6 or vacuum tubes as shown in Fig. 7-7. There are also tuners that use a field-effect transistor (FET) circuit like the one shown in Fig. 7-8. In any of these cases, the basic operation is the same. Fig. 7-9 shows a functional block diagram of an rf amplifier circuit.

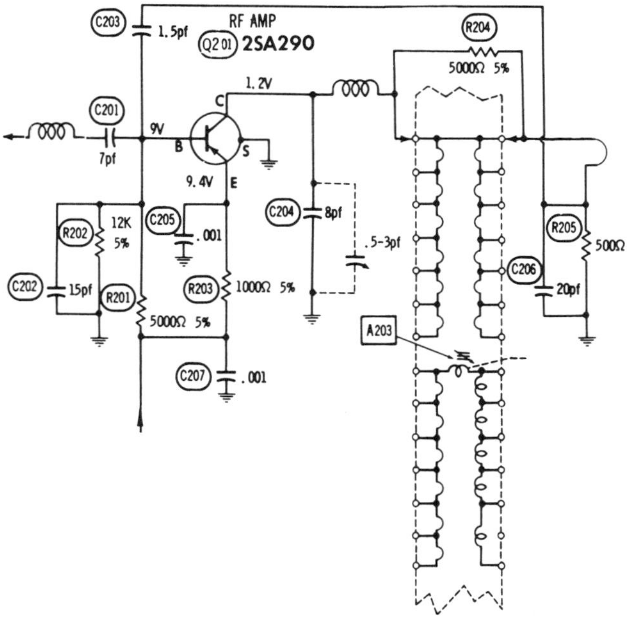

Fig. 7-6. A transistor rf amplifier.

As we have already seen, the amplifying device may be a transistor, a tube, or a FET. The input and output coupling and tuning circuits are switched when channels are changed, and the signal level is controlled by the agc. The agc will be discussed in more detail in the next chapter.

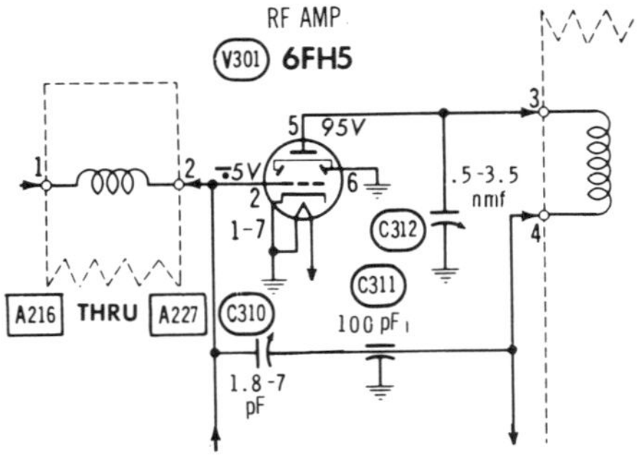

Fig. 7-7. A vacuum-tube rf amplifier.

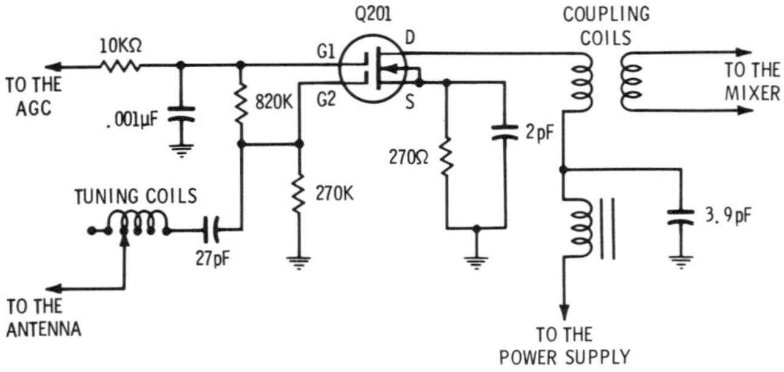

Fig. 7-8. A FET rf-amplifier circuit.

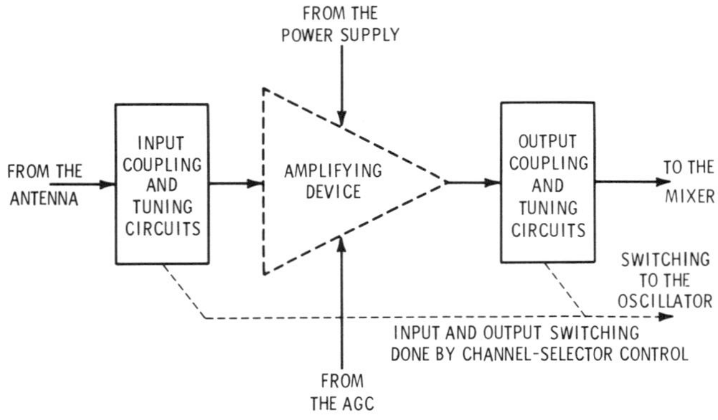

Fig. 7-9. Functional block diagram of an rf amplifier.

7-4 Oscillator Circuits

As you will recall, the oscillator in a television tuner generates a signal that is mixed with the rf-amplifier output signal. To understand better how an oscillator works, examine the circuit shown in Fig. 7-10. For a moment, ignore the dashed lines in the figure. Consider a tuned amplifier circuit like an if or rf amplifier. Such a circuit will have a tuned input and a tuned output as well as an amplifying device like a transistor or a tube. Further suppose that the circuit is tuned to 100 MHz. That is, it amplifies only signals that are at this frequency. If a signal, of say one volt, having a frequency of 100 MHz is injected, then an output will be obtained. The output will be at 100 MHz and will usually be greater than one volt. For example, say that 10 volts is measured at the output.

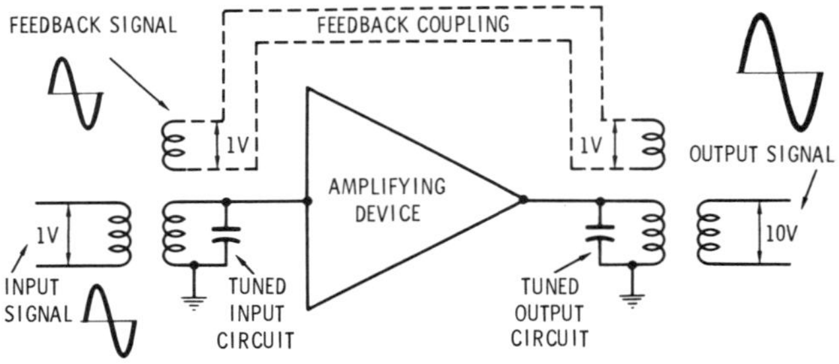

Fig. 7-10. Oscillator action.

Now look at the dashed-line part of the circuit. What do you suppose would happen if a part of the output signal (say 1 volt again) was picked up and taken around to the input and coupled in? Well, you could now take away the original input signal, and the circuit would go on working. In fact, you would not even have to put in an input in the first place. All you have to do is couple the output and input together, and the circuit will produce an output signal all by itself. This is called as *oscillator.*

An oscillator circuit always has three parts. Fig. 7-11 shows these basic parts. The circuit must be able to amplify, so it always has a transistor or a tube. It must be tuned, so it always has a tuned circuit. And finally, it must have a way of getting part of the output back to the input, so it has a feedback circuit.

Now look at some actual oscillator circuits. Fig. 7-12 shows two different circuits. Can you identify the three basc parts of each circuit? In the transistor circuit (Fig. 1-12A), the coil and the 3-pF capacitor are the tuning circuit, and the 1.5-pF

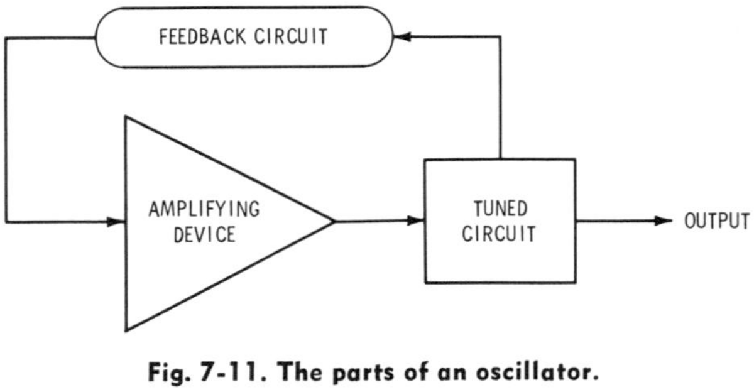

Fig. 7-11. The parts of an oscillator.

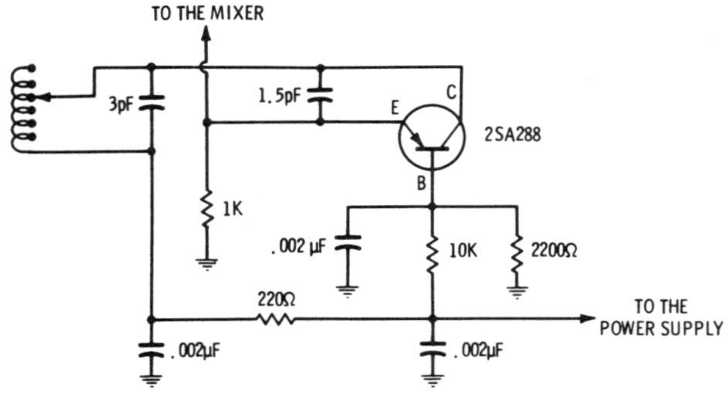

(*A*) *Transistor circuit.*

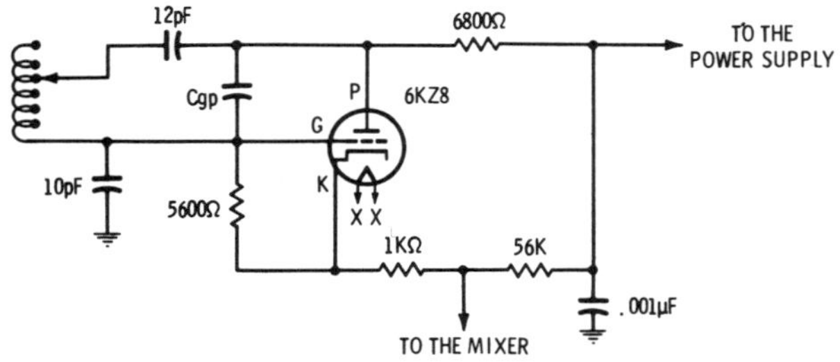

(*B*) *Tube circuit.*

Fig. 7-12. Oscillator circuits.

capacitor is the feedback path. In the vacuum-tube circuit (Fig. 7-12B), the coil, the 12-pF capacitor, the 10-pF capacitor, and the grid-to-plate capacitance (C_{gp}) all form the tuning circuit. Feedback is through these same parts from plate to grid.

7-5 Mixer Circuits

As you would probably expect, the mixer circuits come in the same varieties as do rf amplifiers and oscillators. That is, they may be either solid-state or vacuum-tube circuits. In many cases, the rf-amplifier signal is applied to one electrode and the oscillator signal is applied to another. Fig. 7-13 shows typical examples of a solid-state mixer and a vacuum-tube mixer. Notice that both inputs are applied to the control grid of the vacuum-tube

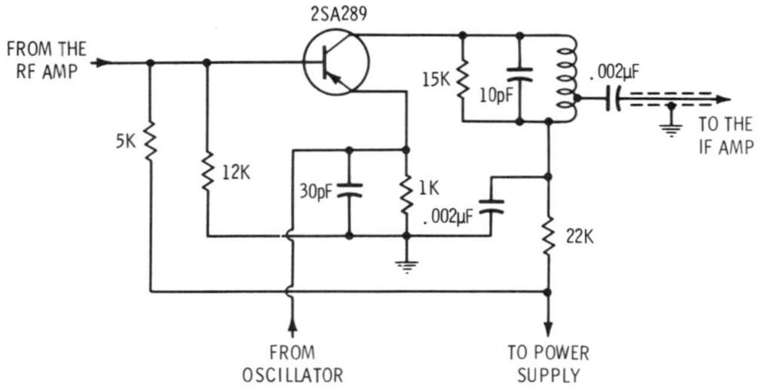

(*A*) *Transistor circuit.*

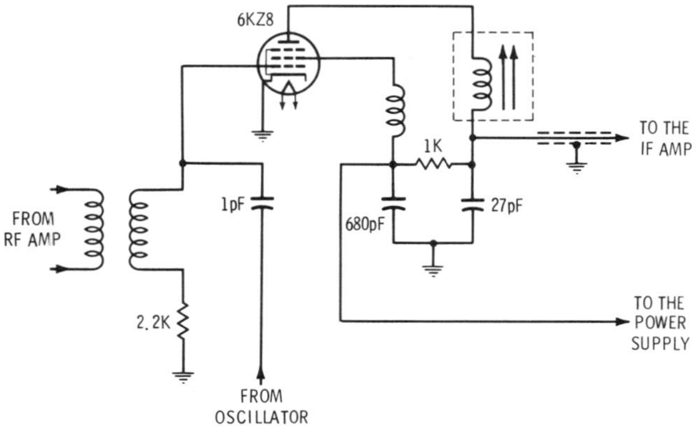

(*B*) *Tube circuit.*

Fig. 7-13. Mixer circuits.

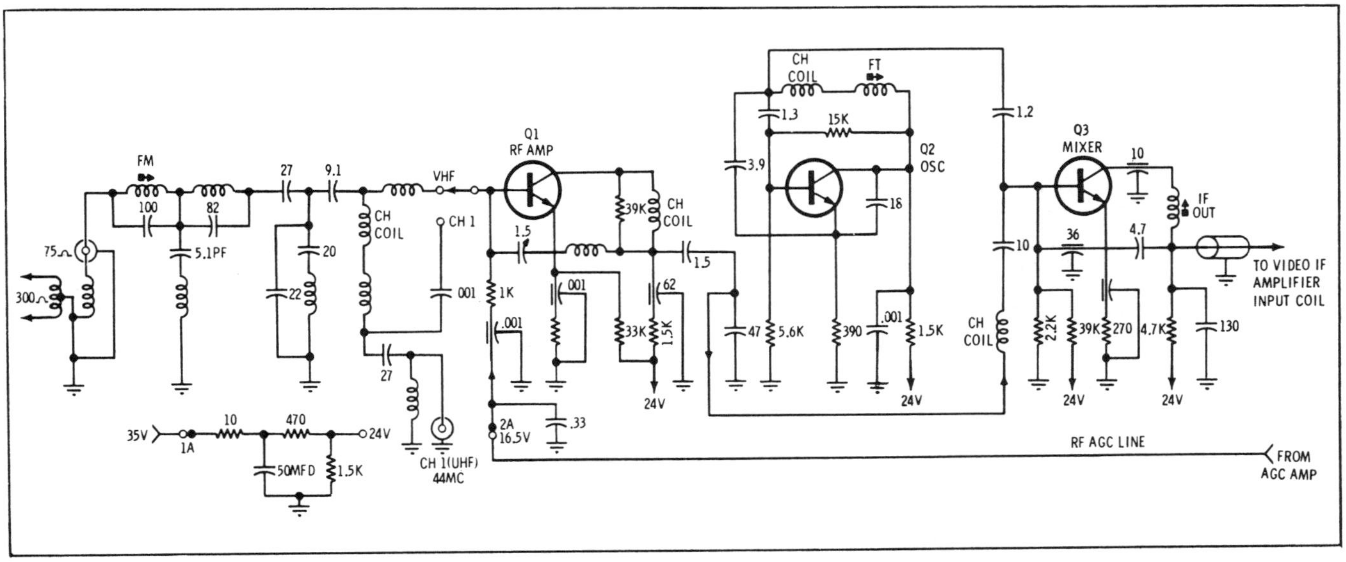

Fig. 7-14. Transistor tuner circuit.

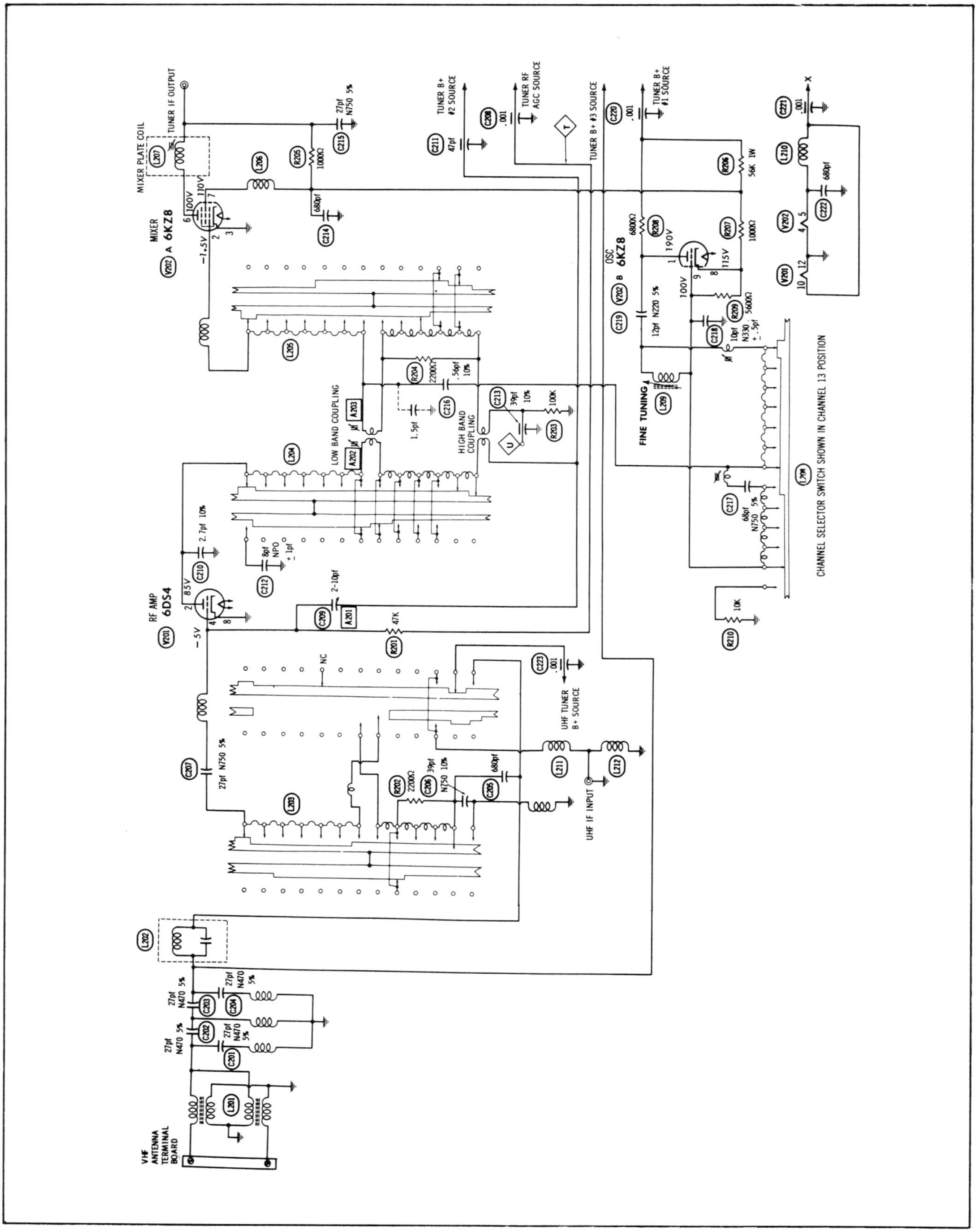

Fig. 7-15. A vacuum-tube tuner circuit.

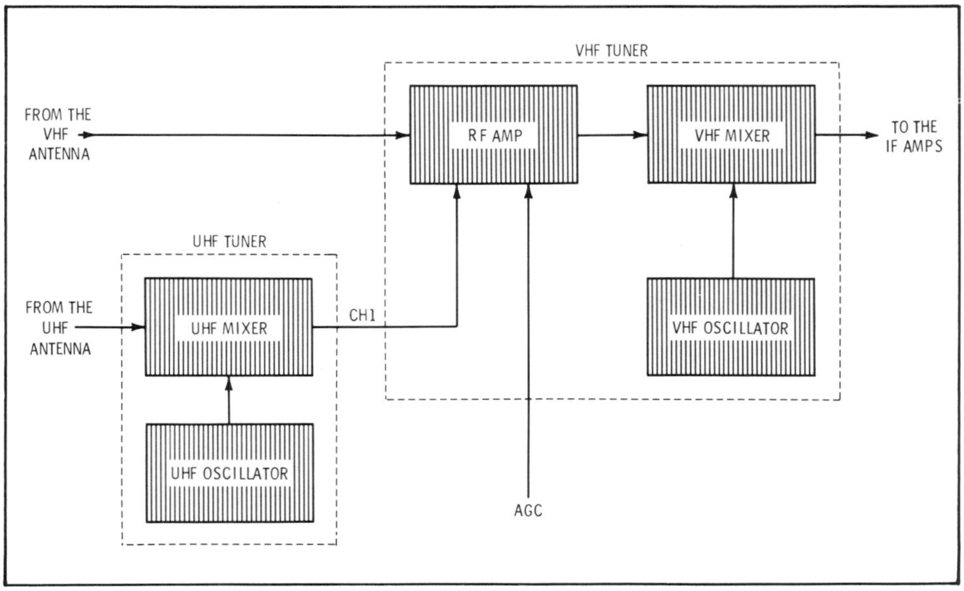

Fig. 7-16. Block diagram of typical tuner stages.

mixer. This type of connection is used in about as many cases as is the separate input one.

7-6 Troubleshooting Inside the Tuner

Troubleshooting tuner circuitry is actually very little different from troubleshooting any other circuit. You should, however, be quite careful in working around the tuner coils. Simply moving the coils can detune the set and cause rather serious problems.

Notice that in the solid-state tuner diagram in Fig. 7-14, the circuit voltages are not shown. Normally these values will be shown on the diagram. However, if they are not, you can still proceed. Notice for example that there is a 1.5K resistor between the 24-volt power supply tap and the collector circuit of the rf amplifier. Because this resistor is present and of a typical collector-circuit value, expect it to drop about half of the supply voltage. Consequently, expect the collector-emitter voltage to be about 12 to 15 volts. If it is *greatly* different from this value, you should suspect the transistor. It is, however, always wise to consult a schematic with voltages shown before making extensive parts tests in a set.

The vacuum-tube tuner shown in Fig. 7-15 is much more typical with respect to the data shown

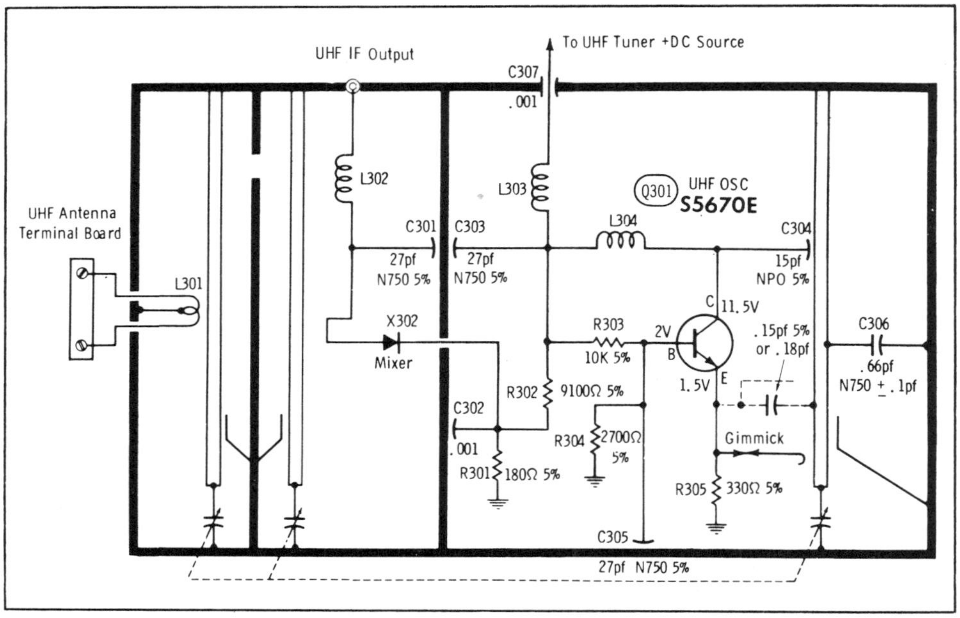

Fig. 7-17. A uhf tuner.

on the diagram. Notice that voltage values are shown for all of the electrodes.

Components in the tuner are checked just as they would be in any other circuit. However, again it is worth mentioning that the tuning coils inside the tuner should *not* be disturbed.

7-7 Ultrahigh-Frequency Tuners

Almost all of the uhf tuners in use today are solid-state ones. You might find an occasional tube-type uhf tuner in an older set. If so, you can usually treat it just like a vhf tuner.

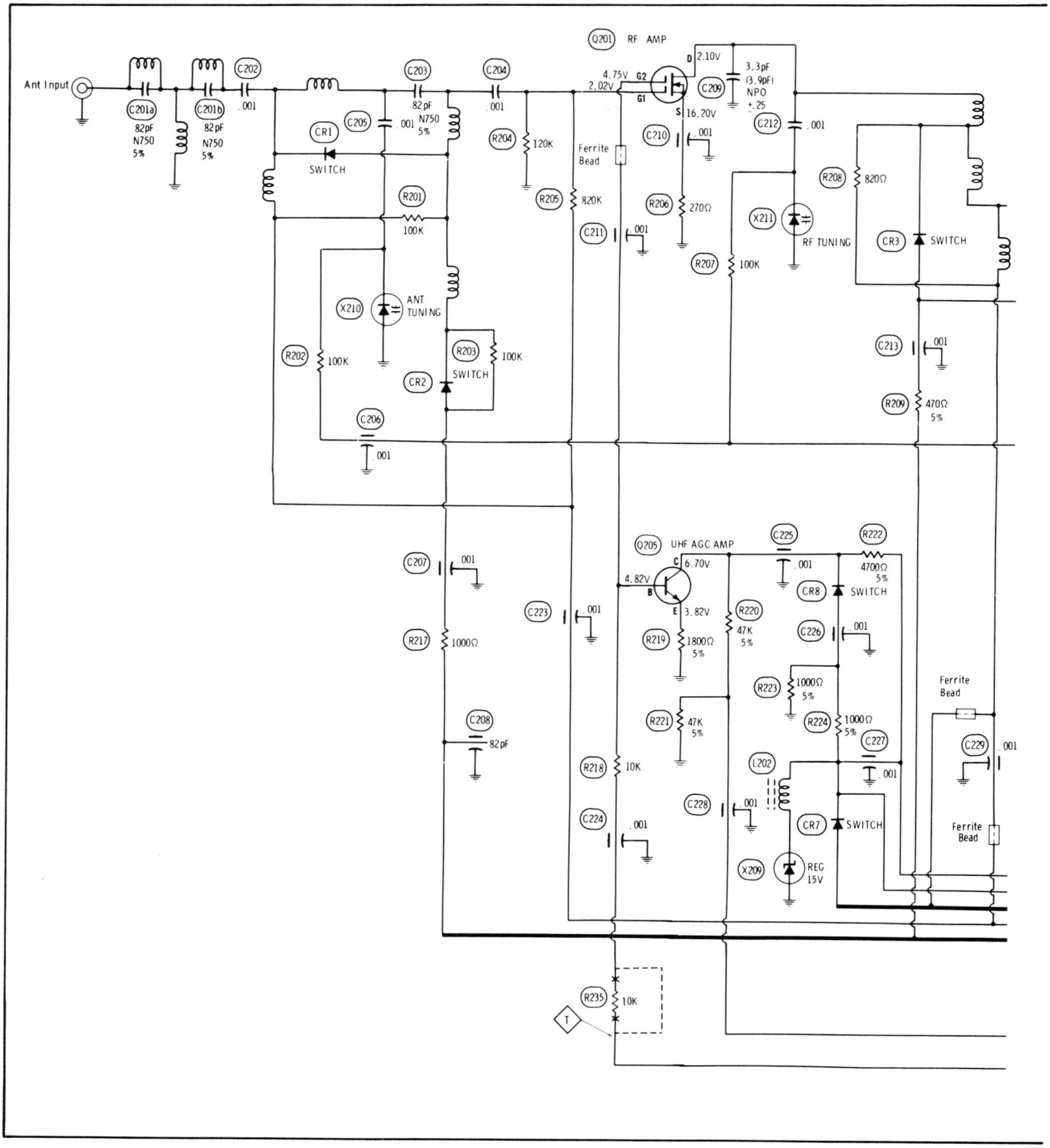

Fig. 7-18. A

In the case of the more recent uhf tuners, you will find them to contain only a mixer and oscillator. When you switch to the uhf position on the channel selector, the uhf tuner is turned on and the vhf circuit is connected differently. The uhf tuner then converts the selected uhf signal so that it can be applied to the rf amplifier in the vhf tuner. Fig. 7-16 shows the interconnection of the tuner stages.

As mentioned before, the uhf tuner contains only a mixer and oscillator circuit. Fig. 7-17 shows a typical tuner of this type. Notice that a

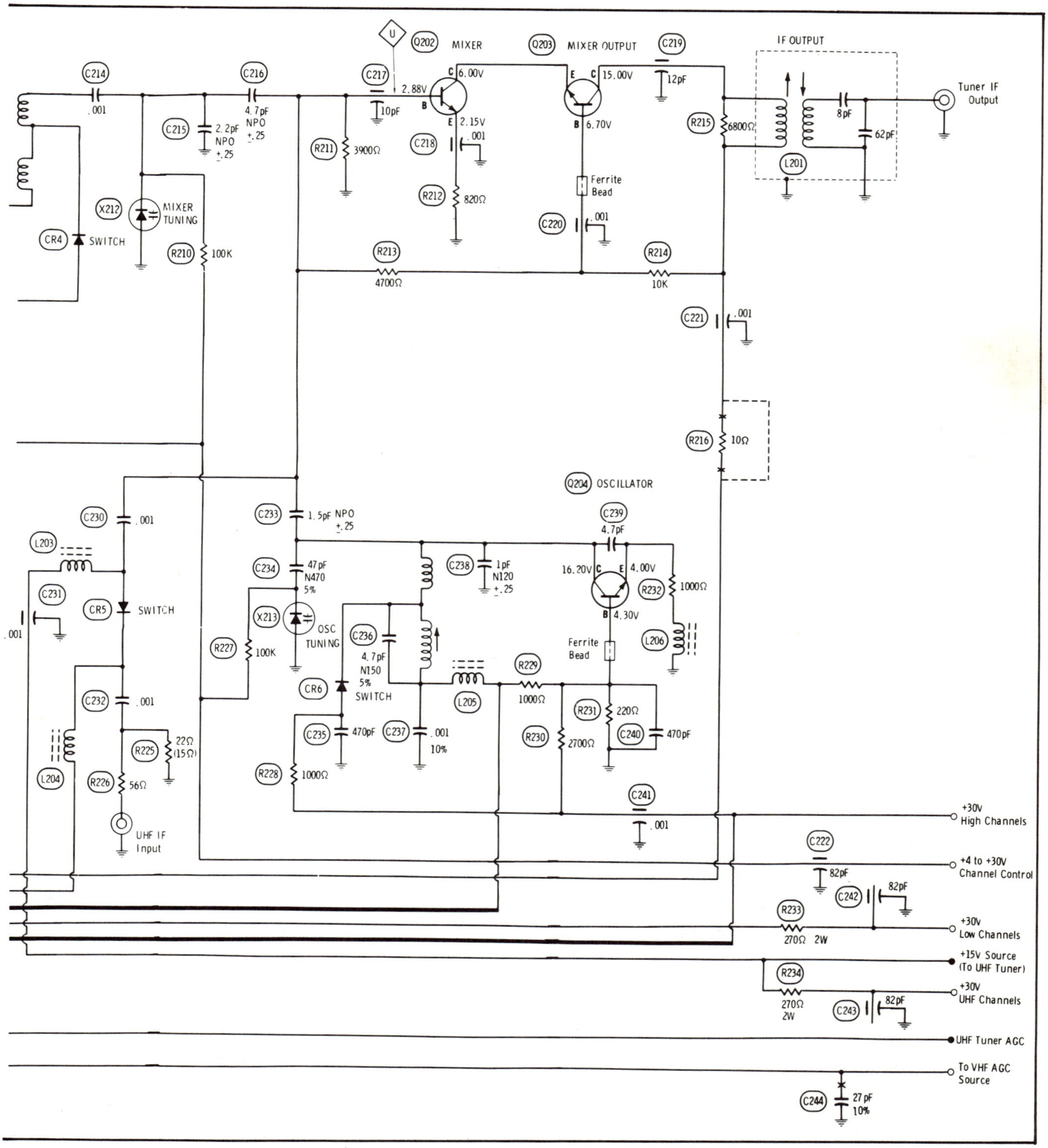

varactor tuner.

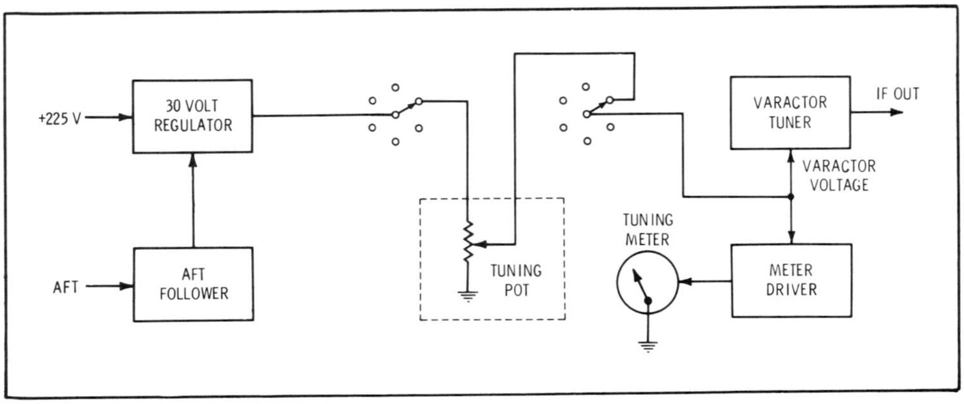

Fig. 7-19. Control system for varactor tuner.

diode is normally used for uhf mixing. The long parallel lines shown in the diagram are tuned transmission lines. They act just like a tuned coil-and-capacitor resonant circuit at ultrahigh frequencies.

Troubleshooting a uhf tuner is quite like any other troubleshooting job. Measure the element voltages, and, using that information, trace down the defective part.

Both uhf and vhf tuners can be rebuilt if they have serious defects. There are a number of companies that specialize in tuner service. You can find out the name of such a company serving your area by asking a local electronic parts dealer.

7-8 Varactor Tuners

In recent years, some tuners have been manufactured that are tuned by electronic means rather than mechanical means. These tuners employ reverse-biased diodes that act as voltage-controlled variable capacitors (varactors). Fig. 7-18 shows the diagram of a typical varactor tuner. Diodes X210, X211, X212, and X213 are the varactors, and they tune the antenna, rf amp, mixer, and oscillator, respectively.

The varactor control voltage is produced by a special control circuit such as the one shown in Fig. 7-19. In this circuit, the set low voltage (225 V) is regulated at approximately 30 volts and is finely adjusted by the automatic fine tuning (aft) circuit. The tuning voltage is then fed through a network of tuning pots to the varactors and the tuning meter.

An advantage of varactor tuning is freedom from problems with contact cleaning and mechanical coupling. Also, varactor tuning allows the tuner to be located in any convenient place in the cabinet rather than on the right front as is the case with mechanical tuners.

7-9 Review Questions

1. Describe the symptoms associated with a defective rf amplifier.
2. The switch in the base circuit of Q1 in Fig. 7-14 is jammed in the Channel 1 position. How would the set act? How could you isolate the defect?
3. How can you tell the difference between a defective mixer and a defective first if stage?
4. The transistor in Fig. 7-6 has a short from collector to emitter. What would be the external set symptoms? How would the circuit voltages change?
5. What symptoms would you expect to see if the oscillator of a television receiver had failed?
6. The 6DS4 tube in Fig. 7-15 is burned out. How would you check it by making circuit voltage measurements? What would be the picture and sound symptoms of this problem?
7. The coil in the plate circuit of Fig. 7-13B is an open circuit. What would be the external set symptoms?
8. What would be the circuit symptoms in Question 7? How would the voltages change?
9. Write out a description of the symptoms that you would expect from a defective mixer stage.
10. The collector-to-emitter voltage of Q2 in Fig. 7-14 is zero. What would you do next?
11. The voltage across the 270-ohm source resistor in Fig. 7-8 is zero. The drain-to-source voltage is zero, but the drain-to-ground voltage is 12 volts. What is the problem?
12. The 1.5-pF capacitor in Fig. 7-12A is an open circuit. What would be the external

set symptoms? Would the circuit voltages change? Explain your answer.

13. The 6KZ8 tube in Fig. 7-15 is burned out. What would the tv screen look like? How would the circuit voltages change?
14. The mixer transistor in Fig. 7-14 is open collector to emitter. What would be the set symptoms? Circuit symptoms?
15. What would be the symptoms of a bad uhf oscillator transistor?
16. The 130-pF capacitor across the output of Fig. 7-14 is shorted. What would be the symptoms seen on the set? How would you isolate the problem?
17. The circuit in Fig. 7-7 has a burned out tube. What would be the external set symptoms? What change would you expect in circuit voltage?
18. A tv set is switched to the uhf position, but the vhf mixer is bad. What would the set do? What would you do next?
19. What symptoms would you expect if the mixer diode in a uhf tuner were shorted?
20. The coil in the collector circuit of Fig. 7-13A is open-circuited. How would this affect the set operation?
21. How would you start to solve the problem in Question 20?

CHAPTER **8**

Automatic Gain Control

When a television set is switched from one station to another, we would like the picture (and sound) quality to stay the same. If all television stations delivered the same signal strength to the set, the problem of uniform quality would not be so difficult to deal with.

Unfortunately, different stations have different output levels. Also, the signals from different stations come to us over different distances and from different directions. The signals may be affected by different weather conditions and many other factors. As a result, two different stations will almost always provide different signal levels to the receiver. In addition, many of the factors that affect the received signal will change from minute to minute.

8-1 Purpose of Automatic Gain Control

The *automatic gain control* (agc) section adjusts the set to correct for these variations in the received signal. To do this, the agc circuit samples the received signal at the video amplifier as indicated in Fig. 8-1. If the video signal is too weak, the agc circuit increases the amplification of the if amplifiers and the rf amplifier. If the video signal is too strong, the agc decreases the rf and if amplification. In this way, the agc system continuously adjusts the receiver circuits to produce a good picture in spite of the variations in the incoming signal strength.

Inside the agc block, two things will usually be found. They are:

1. An agc keying circuit, which is sometimes called the *agc gate*.
2. An amplifier for the agc signal.

Fig. 8-2 shows the interconnection of these two circuits.

8-2 Producing the Automatic Gain Control Signal

As the television signal at the antenna of a receiver increases, the video level also increases. This increase in video level is felt at the first video amplifier, and the signal at the video amplifier is coupled to the agc keying circuit shown in Fig. 8-3. The agc keying transistor (Q1) is connected so that it is completely shut off except during the time that it gets the positive pulse from the horizontal-sweep circuit. Since these pulses come 15,750 times a second, the keying circuit samples the video that many times every second. These many little video samples are averaged out by the 10-μF capacitor and the 5.6K resistor. This average agc voltage is sent on to the agc amplifier circuit.

The diode (D1) in Fig. 8-3 prevents negative pulses from discharging the capacitor through the keying transistor.

Some sets do not have an agc amplifier circuit. In such a case, the output agc voltage from the keying circuits is applied directly to the if and rf amplifiers. In sets that use an integrated-circuit if section, the agc signal may be picked off at the if output of the IC rather than from the video section.

8-3 The AGC Amplifier

Many sets have an agc amplifier stage following the agc keying circuit. Fig. 8-4 shows the

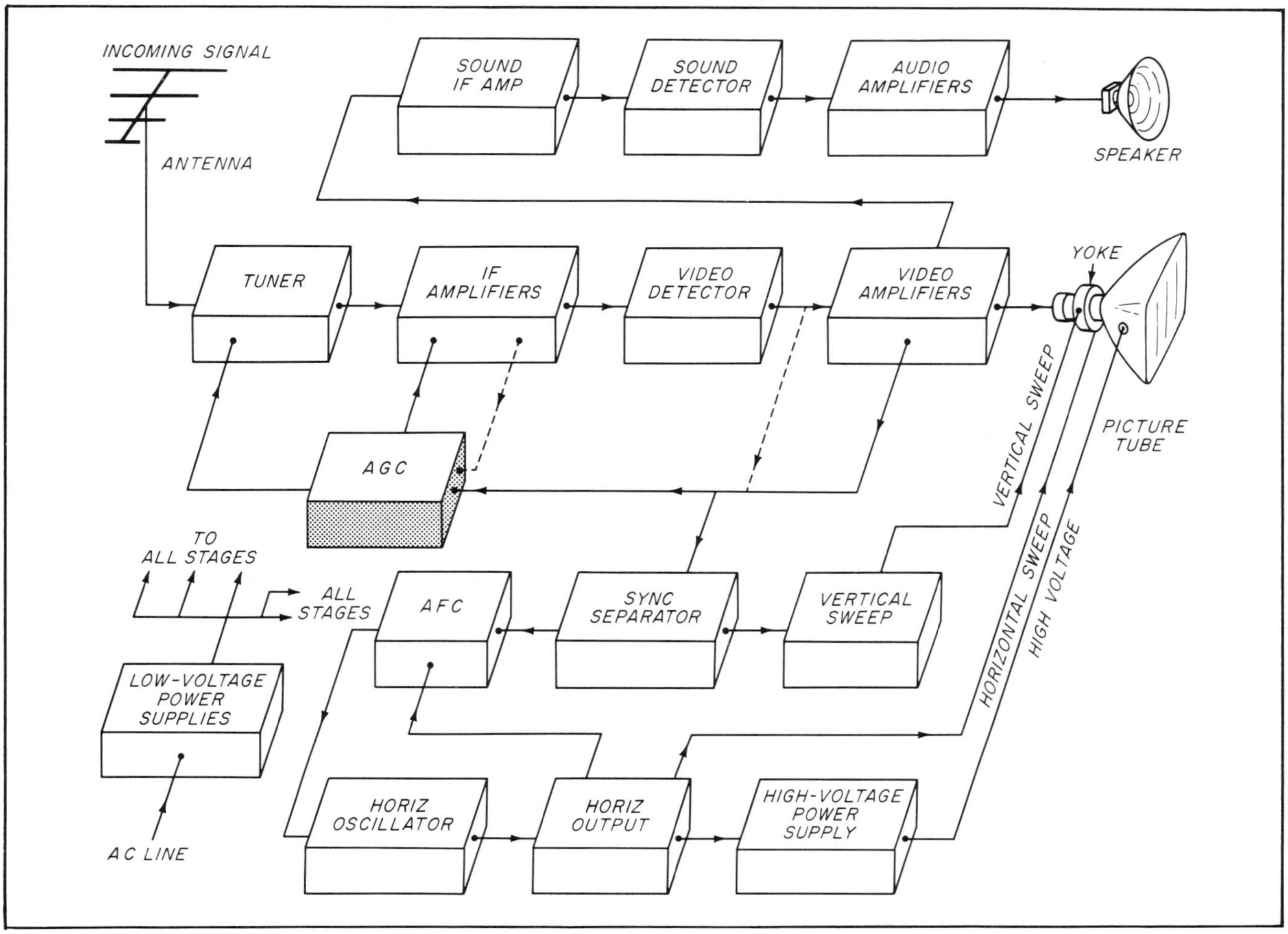

Fig. 8-1. Block diagram of a typical black-and-white receiver.

circuit diagram of such an amplifier. At the output of this circuit, there are both of the basic types of agc. Reverse agc tends to reduce the collector current of the transistor being controlled (the if amplifier in this case). Forward agc tends to increase the collector current of the transistor being controlled (the rf amplifier in this case). It is very important to understand that both forward and reverse agc *reduce* the amplification of the stage being controlled. Consequently, both the if and rf amplifier act to reduce the signal in the set when agc is applied. A few sets use only reverse agc. Others use only forward agc, and some use both types.

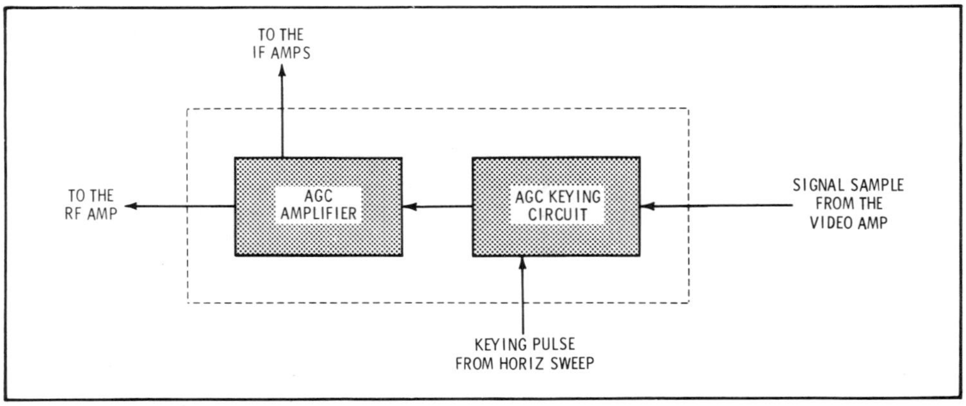

Fig. 8-2. Typical agc circuit.

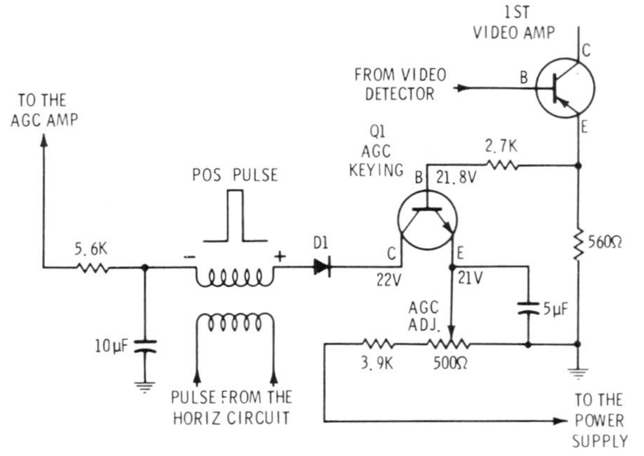

Fig. 8-3. An agc keying circuit.

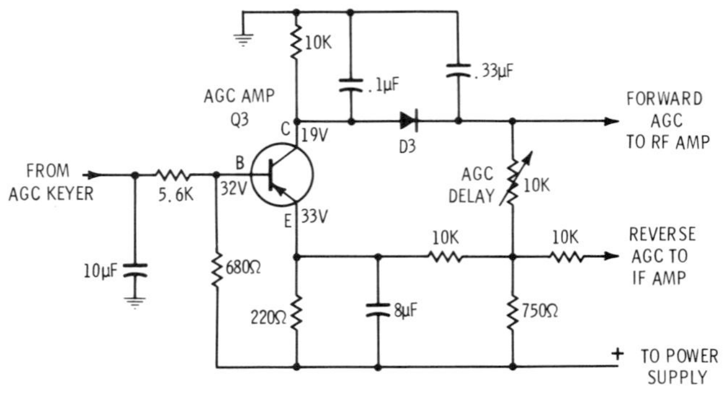

Fig. 8-4. An agc amplifier circuit.

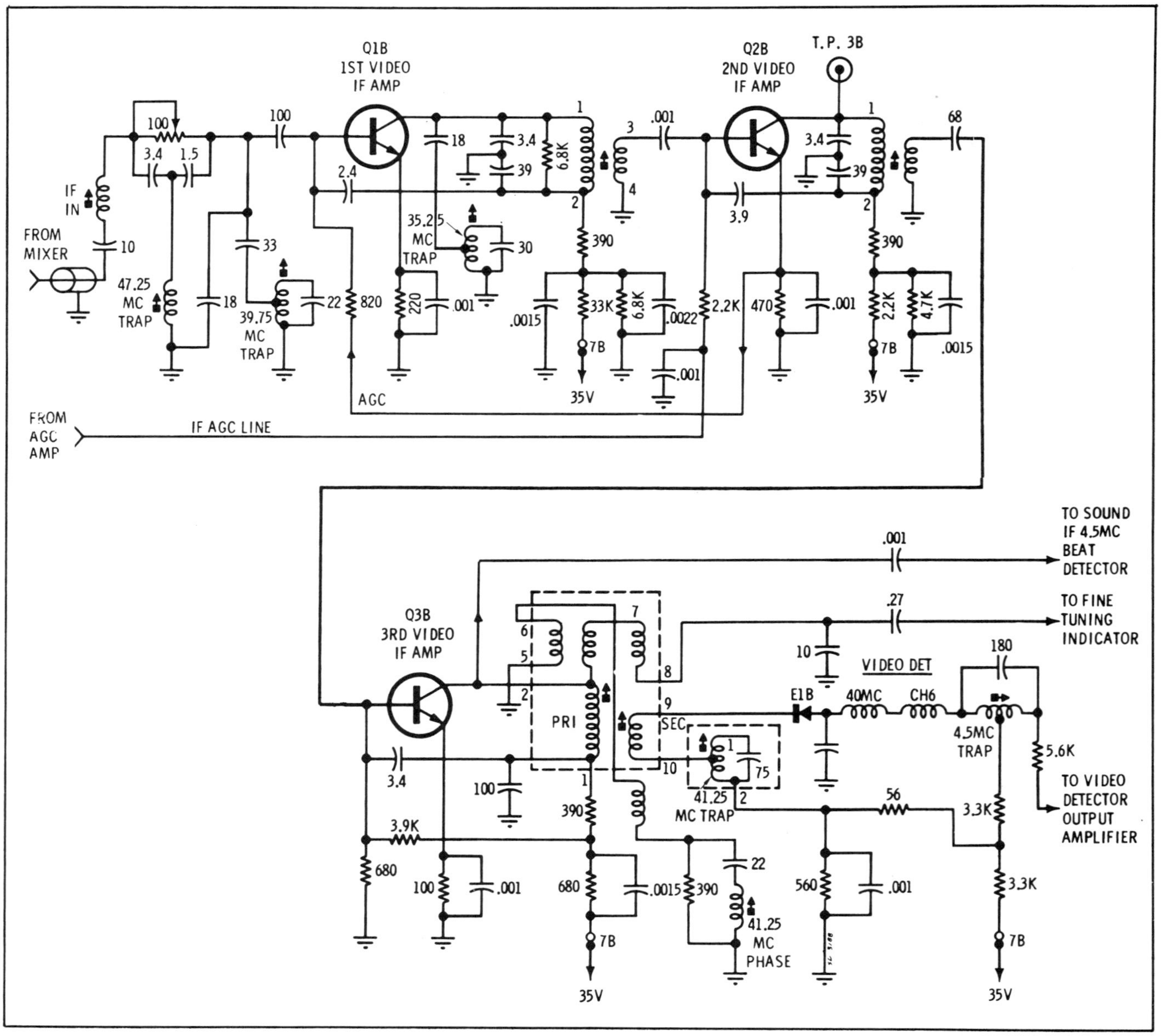

Fig. 8-5. Connection of agc to an if section.

8-4 AGC Action in the IF Amplifier

Either forward or reverse agc may be applied to the picture if amplifier. However, since reverse agc tends to turn a transistor completely off very suddenly but forward agc is more gradual, the forward type is more common.

The agc could be applied to the collector, base, or emitter circuit of the if amplifier transistor. The base-circuit control is usually preferred because it requires very little power from the agc circuit. Fig. 8-5 shows how agc can be applied to the base circuit of an if amplifier.

Notice that in this set agc is first applied to the second if stage. It is then picked up again at the emitter of the second if and applied to the base of the first if amplifier stage. Although many sets use agc only in the first if stage, controlling both of the first two stages provides more gradual control and perhaps a wider range of control.

Fig. 8-6 shows a vacuum-tube if amplifier section with agc applied only to the first if stage. While it would be possible to devise other agc arrangements for a vacuum-tube circuit, current sets use only reverse agc applied to the control grid of the if amplifier.

8-5 AGC Action in the RF Amplifier

Automatic gain control is normally applied to the rf amplifier in the tuner as well as to the if amplifiers. However, to get the best performance in terms of amplification and signal-to-noise ratio, the rf amplifier is usually operated at nearly full gain for all but the strongest signals.

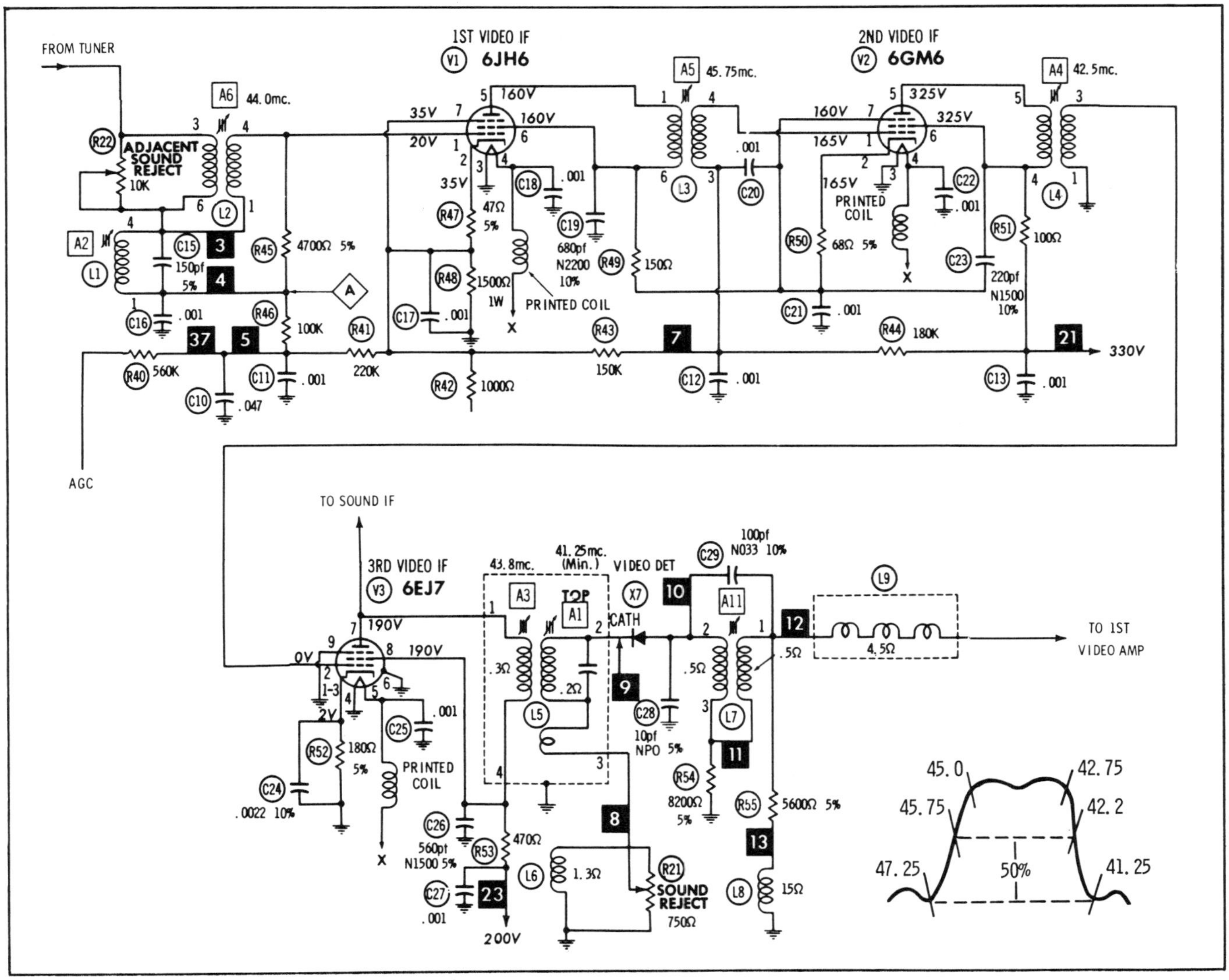

Fig. 8-6. A vacuum-tube if section.

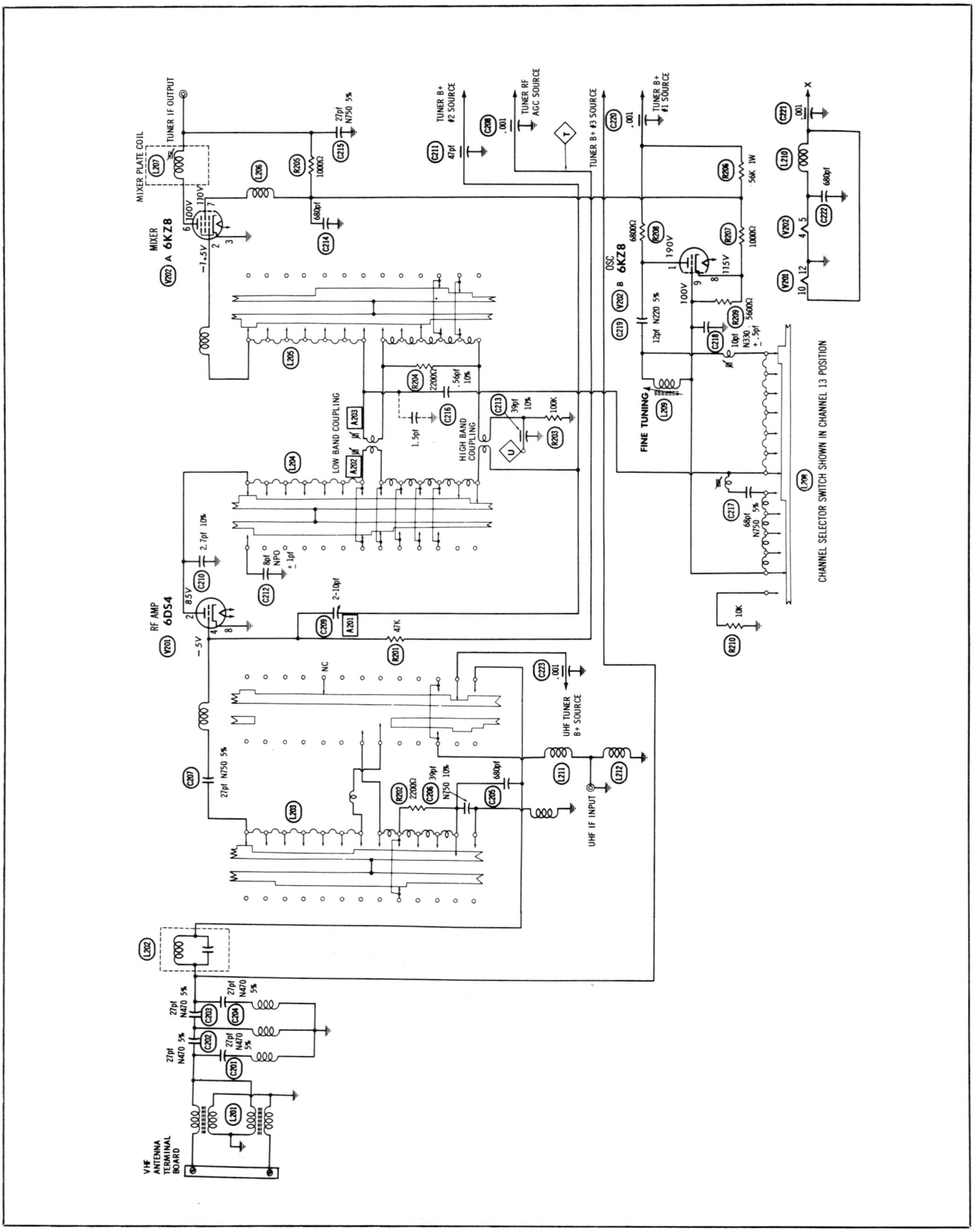

Fig. 8-7. Connection of agc to a vacuum-tube tuner.

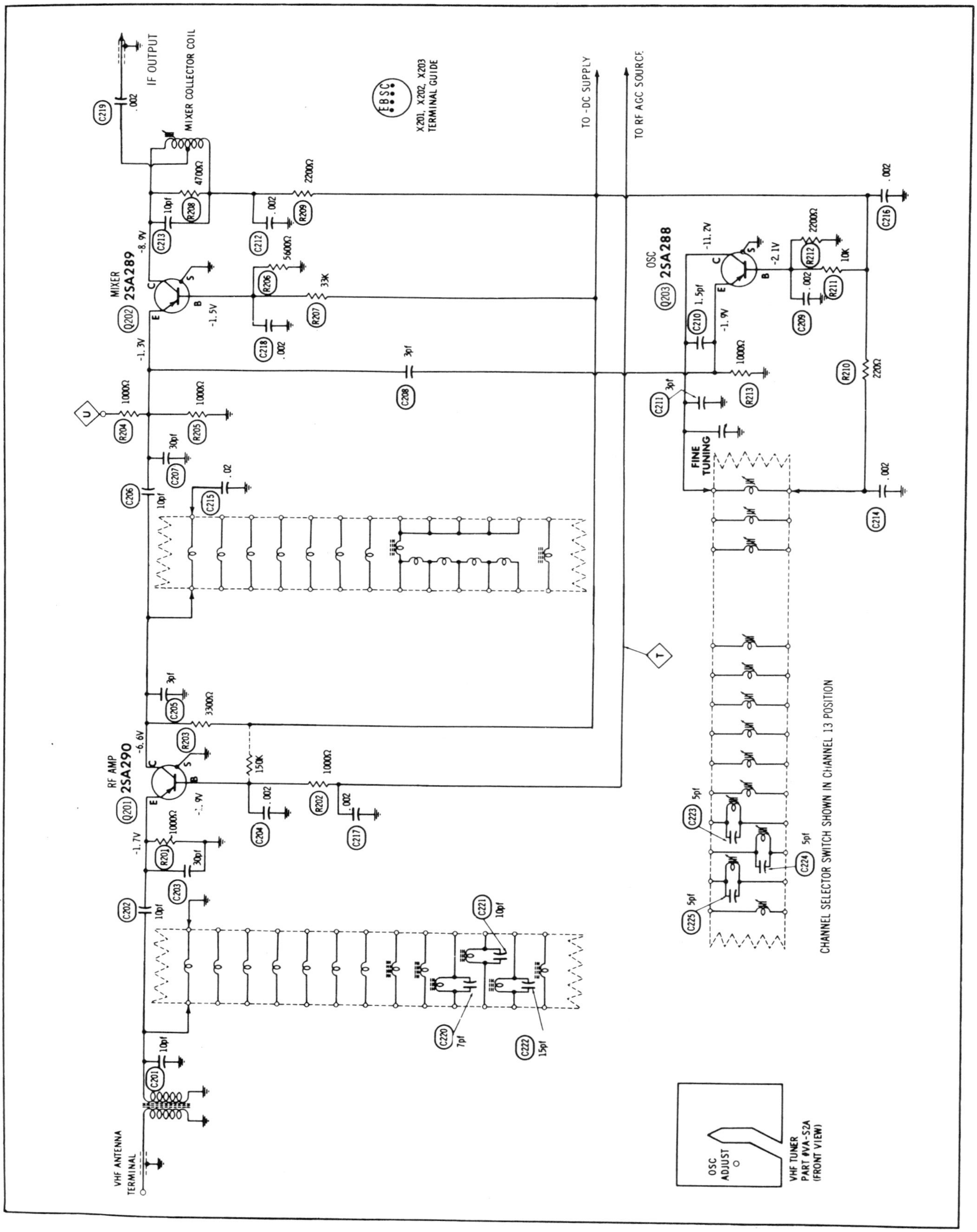

Fig. 8-8. Connection of agc to a solid-state tuner.

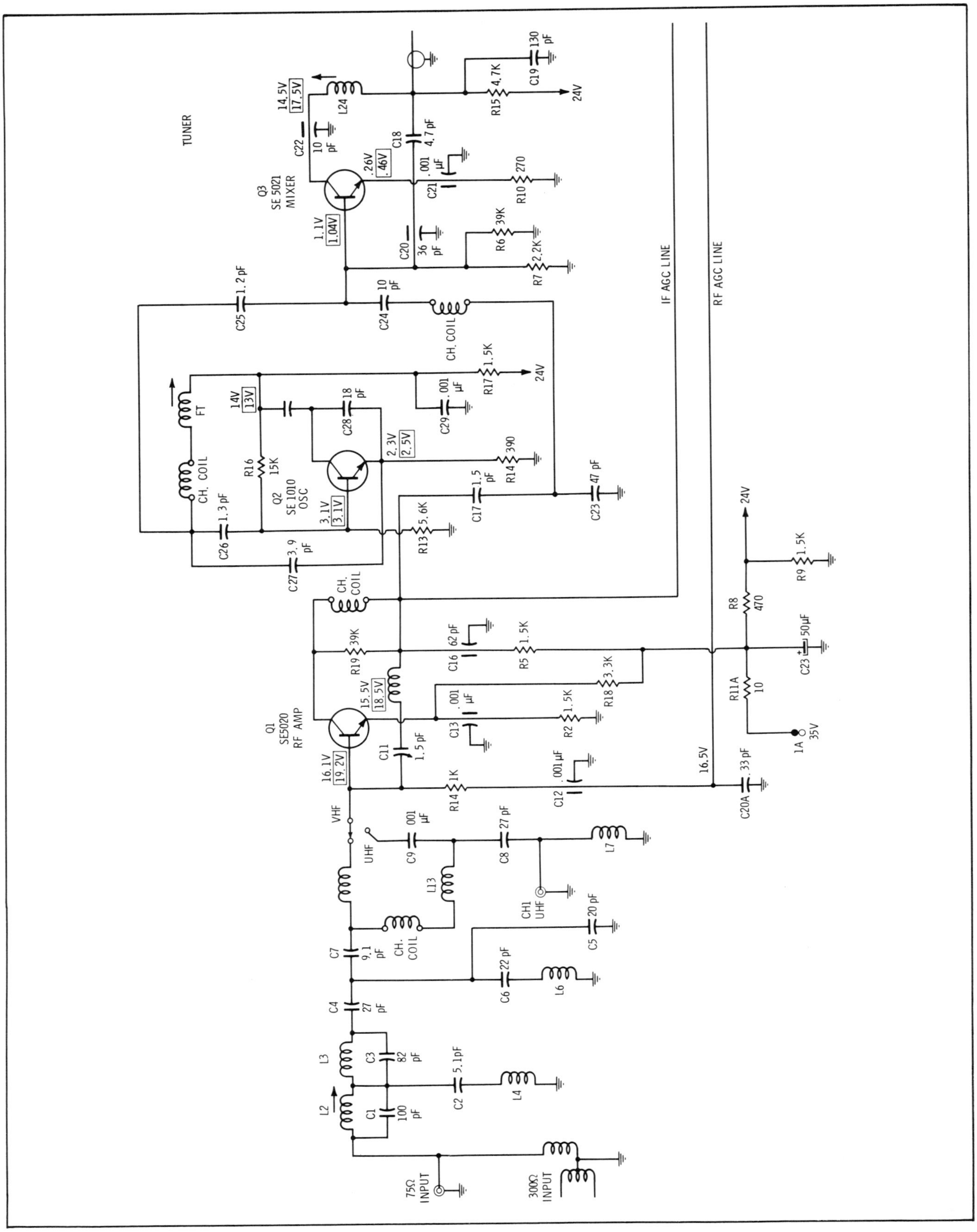

Fig. 8-9. Double tuner agc.

For weaker signals, the if amplifier does nearly all the controlling. Holding off the tuner agc until the signal gets large is called *delayed agc*. Most sets today employ delayed agc in the tuner.

Fig. 8-7 shows a typical agc connection to a vacuum-tube tuner. Notice that the agc is applied only to the control grid of the rf amplifier. Tube-type tuners almost always use reverse agc.

In a solid-state tuner such as the one shown in Fig. 8-8, the agc is usually applied only to the base of the rf amplifier transistor. The agc system used in this type of circuit can be either forward or reverse agc, but the forward type is much more common.

In some sets, both delayed and regular agc are used in the tuner. Fig. 8-9 shows one way in which this can be done. In this circuit, the delayed agc is applied directly to the base of the rf amplifier as before. In addition, the if agc line is coupled into the collector circuit of the rf amplifier. Such a system would tend to give the tuner slightly more control of the medium-level signals, and, of course, increase the range of tuner control.

Some sets have fixed delay in the tuner agc; that is, the amount of agc is fixed at manufacture. Other sets include a control that is used to adjust the delay over a limited range. Such a delay control is often called a *noise* adjustment.

8-6 AGC Trouble Symptoms

Since the agc system controls the amplification, or gain, of the set, agc failures cause too much or too little signal strength. Probably the most dramatic agc symptoms are observed when the if amplifiers have too much amplification. Fig. 8-10 shows a severe case of picture distortion caused by excessive if amplifier signal. This symptom is called a *negative picture*, because the parts of the picture that should be black are white and vice versa. This symptom is more common in tube and hybrid sets than it is in all-transistor sets. If the defect is less severe, the picture will have too much contrast and may bend. Fig. 8-11 shows this symptom. There is usually a raspy buzz in the sound. Varying the contrast control will have only little effect on the picture.

Fig. 8-10. A negative picture.

Fig. 8-11. Bent picture with excessive contrast.

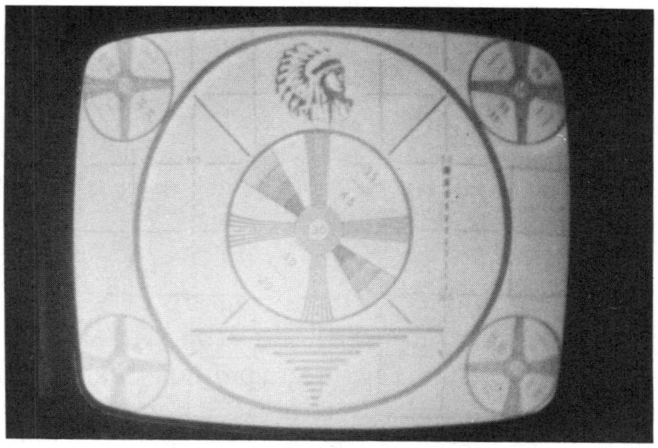

Fig. 8-12. A faded picture.

Both of the foregoing symptoms were caused by too much amplification in the set. What would the screen look like if there was not enough amplification? Fig. 8-12 shows a washed-out screen with weak video caused by too much agc (not enough amplification). This symptom could be caused by a bad if amplifier or a bad video amplifier. In fact, that is why many people think that agc troubles are hard to spot.

It is possible to have an agc problem that completely cuts off the if amplifiers. The result of such a failure is a white screen like the one shown in Fig. 8-13. This problem, of course, looks exactly like an if failure, and it easily could be mistaken for a video-amplifier trouble.

Fig. 8-13. The if amplifier cut off by excessive agc.

8-7 Troubleshooting AGC Problems

As mentioned before, some agc trouble symptoms are hard to separate from picture-if and video-amplifier troubles. What is needed is a way to recognize an agc problem.

A quick way to check for agc trouble is to short-circuit the if agc line and watch the screen. If the agc is working, you should see quite a change in picture contrast.

If agc trouble is suspected, try this test: Hook a voltmeter from the if agc line to ground. Then switch the set from a strong station to a channel with no station. The voltmeter reading should change several volts if the agc is working.

Remember, if the if amplifiers, the detector, or the first video amplifier is dead, there will not be any agc voltage.

When an agc trouble has been identified by using one of the tests above, you troubleshoot the circuits just as you would any other. That is, you measure circuit voltages and compare them to the values given on the diagram of the set.

If it is uncertain whether the agc is working or not, take a battery of about 3 or 4 volts, and hook it from the agc line to ground. In doing this, be sure to check the diagram to get the polarity right. With the agc voltage "fixed" in this way, the rest of the set can be checked out. If the set works with the agc voltage fixed but not without it, then there is an agc problem. On the other hand, if the set does not work with fixed agc (be sure the polarity is right), then there is an if or video problem.

8-8 Review Questions

1. There are circuit diagrams of several sets in the back of this book. Examine each circuit and answer the following questions:
   A. Does the set have an agc keying circuit?
   B. Does the set have an agc amplifier?
   C. Does the set have delayed agc to the tuner?
   D. Is agc applied to more than one if stage?
2. Describe what a *negative picture* looks like to you.
3. What would be the symptoms if the delayed agc cut the rf amplifier completely off when a very small signal was received?
4. Why do we seem to worry more about if agc failures than we do about tuner agc failure?
5. What would be the external set symptoms if transistor Q1 in Fig. 8 - 3 had an open circuit between the collector and emitter?
6. How would the problem in Question 5 affect the agc voltage?
7. What would be the external set symptoms if the 8-μF capacitor in Fig. 8-4 were shorted?
8. How would the trouble in Question 7 affect the if agc voltage? The rf agc voltage?
9. A set with the if section shown in Fig. 8-6 seems to have no agc (the picture is contrasty and bent). Checking the circuit, we measure zero volts at test point *A*, but the agc voltage is -6 volts just to the left of R40. Which components would you suspect?
10. The agc voltage at test point *T* in Fig. 8-8 is zero on all channels. What would you do next?
11. What set symptoms would you expect if the agc amplifier in a set were completely burned out?
12. Explain in your own words how you determine if the symptoms shown in Fig. 8-12 are caused by an agc problem.
13. Can you tell by looking at the screen whether the symptoms in Fig. 8-13 are caused by an agc trouble?
14. What would you do if you were servicing a set that looked like Fig. 8-13 on all channels?

CHAPTER 9

Troubleshooting With an Oscilloscope

An oscilloscope is nothing more than another type of voltmeter. It allows us to measure voltages, and at the same time to see how the voltage is changing. Take a closer look at the problem of measuring a changing voltage. Suppose that you have a voltage that is changing in the manner shown in Fig. 9-1. To start with, suppose that the time units along the horizontal line are minutes. An ordinary voltmeter connected across the circuit that has this voltage would start out reading 15 volts and rise slowly until, after 6 minutes, it would be at 30 volts. Then the voltmeter reading would drop back to 15 volts in 2 minutes time. The voltmeter reading would repeat this 8-minute cycle over and over. Since the voltmeter reading is changing very slowly, you could get a good idea of how the voltage was changing as time passes.

Suppose that the time units in Fig. 9-1 are in seconds rather than minutes. This being the case, the voltmeter reading would rise from 15 to 30 volts in 6 seconds. Then it would drop back to 15 volts in 2 seconds. These changes are occurring fast, but you would still get a fair idea of how the voltage is changing with time.

If the time units in Fig. 9-1 are changed to tenths of seconds, you have a real problem. The voltage would go through its whole cycle in 8 tenths (0.8) of a second. Even if the voltmeter could follow such a rapid change, you could not tell much about the changes by watching it. You could tell, however, that it was changing.

In a television set, you do have a voltage that changes in much the same manner as the voltage shown in Fig. 9-1, but it goes through the whole cycle 60 times every second. If a voltmeter is connected across a voltage changing that fast, it will not be able to keep up with the changes. In fact, the voltmeter would read a constant value somewhere between 15 and 30 volts (about 24 volts).

These very rapid changes in voltage are extremely important to the operation of a television set. What is needed is a device that can follow these changes and allow measurement of the voltages. An oscilloscope can do this.

9-1 Oscilloscope Cathode-Ray Tubes

An oscilloscope is really a lot like a television set, but much simpler. As in a tv set, the main part of a *scope* (the word "scope" is used as a short form for "oscilloscope") is a cathode-ray tube (crt). Fig. 9-2 shows a sketch of the type of crt used in a scope.

The electron gun and the high-voltage anode in the scope tube are just like those in a tv tube. The screen is made the same way, but the phosphor material is different. The deflection system in the scope tube is quite different from that of a tv picture tube. In the scope tube, electrostatic deflec-

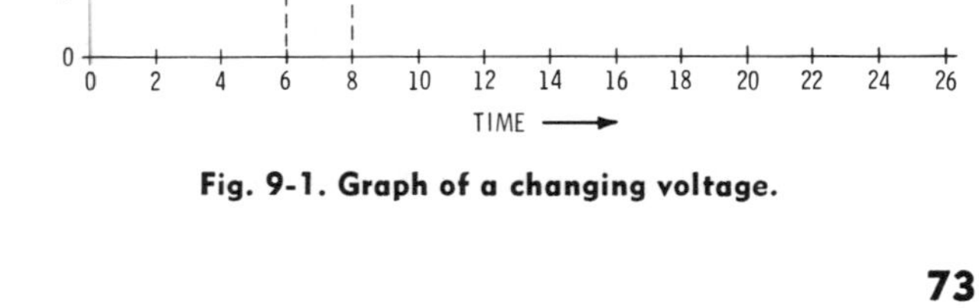

Fig. 9-1. Graph of a changing voltage.

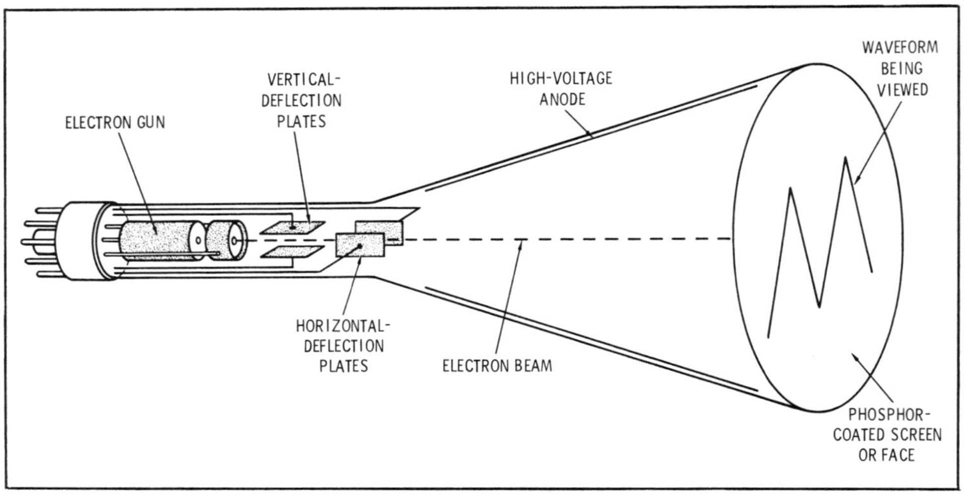

Fig. 9-2. An oscilloscope crt.

tion plates are used. If a voltage is applied across the vertical-deflection plates, the electron beam will be bent toward the positive plate and away from the negative one. In this way, the electron beam is caused to move up and down in step with the vertical-deflection voltage. When the beam moves, so does the spot of light where the beam strikes the screen. By applying a voltage across the horizontal-deflection plates, the spot can be moved horizontally.

9-2 Oscilloscope Operation

Working with the crt in an oscilloscope are three other basic systems. Fig. 9-3 shows these systems in block-diagram form. Fig. 9-4 shows the basic controls found on the front panel of a typical scope.

The power supplies provide the required operating voltages for all sections of the scope. They also contain controls for adjusting the *intensity* (brightness) and *focus* (sharpness) of the spot.

The horizontal-sweep generator is connected to the horizontal-deflection plates. This generator produces a voltage that moves the spot from left to right at a constant rate. When the spot reaches the right side of the screen, the sweep generator returns it very quickly to the left side. Then it *sweeps* it across the screen from left to right again.

As a result of the sweep-generator action, a bright horizontal line is seen on the face of the scope tube. The rate at which the spot moves from

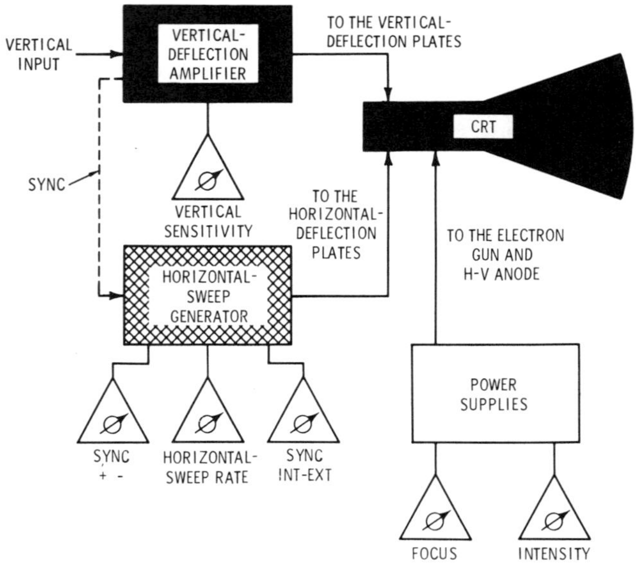

Fig. 9-3. Block diagram of an oscilloscope.

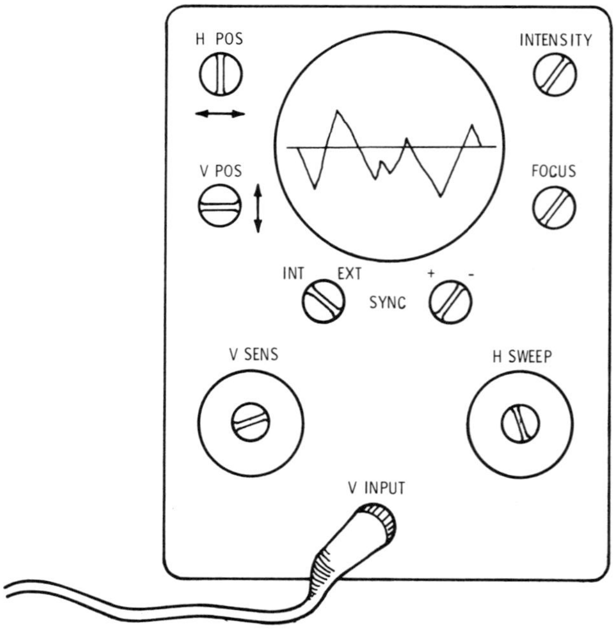

Fig. 9-4. Simplified controls on an oscilloscope.

left to right can be adjusted with the sweep-rate adjustment. Also located in the horizontal-sweep generator, but not shown in Fig. 9-3, is a *horizontal-position* control. This control allows the whole horizontal line to be moved back and forth.

The vertical-deflection amplifier simply takes the signal we wish to look at, amplifies it, then applies it to the vertical-deflection plates. In this amplifier is the *vertical-sensitivity control.* This control is like the range switch on a voltmeter. It sets the amount of voltage required to just fill the screen vertically. There is also a *vertical-position control.* This control allows the whole sweep to be moved up and down.

Now, if the crt spot is moving up and down in step with the voltage connected to the vertical input and from left to right in step with some units of time, the voltage changes with time can be seen. Since the sweep rate can be controlled, the display can be adjusted to show very rapidy changing voltages.

There is one more point that should be considered—synchronization. It would be very convenient if the scope would start its horizontal sweep just when the vertical voltage starts its cycle. On most scopes, there are two *sync* controls. One is a switch that has two positions—internal and external sync. On *internal sync,* the horizontal sweep is started by the signal being viewed. On *external sync,* you can put in any signal you want to sync to, and then view another signal. There is a connection on the scope for hooking up this external sync.

The other sync control is usually a dial that is marked + at one end and − at the other. If you set it to +, the scope syncs on the signal when it starts going *up;* on − sync, the scope starts its sweep when the signal starts going *down.*

9-3 Adjusting the Scope for Waveform Viewing

To set up the oscilloscope so that you can look at a waveform, follow the steps given below:

1. Turn on the scope and allow it to warm up for several minutes.
2. Set the sweep-range selector switch to a position in the middle of its range.
3. Turn the intensity control up to the middle of its range.
4. Adjust the vertical- and horizontal-position controls until you have a horizontal line across the middle of the screen.
5. Adjust focus and intensity for a sharp clear line that is not too bright.
6. Connect the voltage to be viewed to the vertical input.
7. Set the sync for internal +.
8. Adjust the vertical sensitivity until the display is about one-half the height of the screen.
9. Adjust the horizontal-sweep rate until you have a single trace showing the desired number of cycles.

The procedure given above is a general one that will work for most scopes. Even so, it is wise to read the operation manual for your scope before you attempt to use it.

9-4 Waveforms Found in a TV Set

In a television receiver, there is a large variety of waveforms that can be viewed. There are some that are present and cannot usually be viewed. Some of the ones that cannot normally be viewed are:

1. The picture-tube high voltage. It will damage the scope.
2. The horizontal-output pulses. They will damage most scopes.
3. The signal in the tuner. Most scopes cannot show signals of such high frequency.
4. The picture if signal. Same reason as for signals in the tuner.

Some of the waveforms that can be viewed in a television receiver are of relatively little practical interest. Among these are:

1. The input ac power. It is simply a sine wave.
2. The audio signal. Listening to it is normally more useful.
3. The agc voltage. It does not change fast enough—a voltmeter is better.
4. Most dc operating voltages. Same reason as for agc.

There are other waveforms that can be viewed and which contain considerable useful servicing information. Perhaps the most important of these are:

1. Video
2. Sync pulses

3. Vertical-sweep signals
4. Horizontal-sweep signals
5. Agc keying pulses
6. Color signals
7. Sometimes the power-supply ripple

Emphasis should be placed on this last group. Fig. 9-5A shows photographs of several television waveforms from this group. Fig. 9-5B shows the points in the tv set where these waveforms may be found.

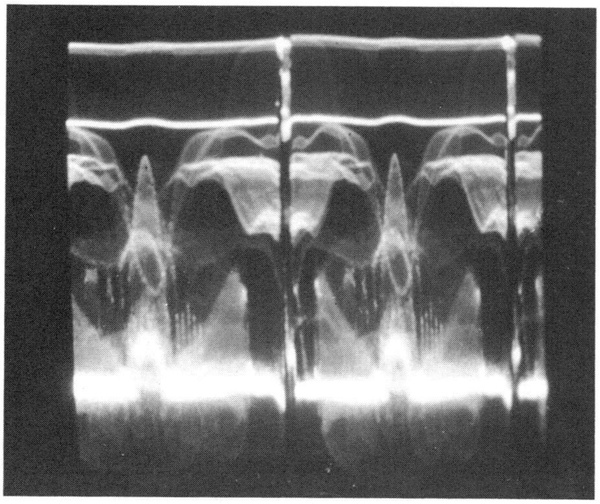

(1) VIDEO WAVEFORM.

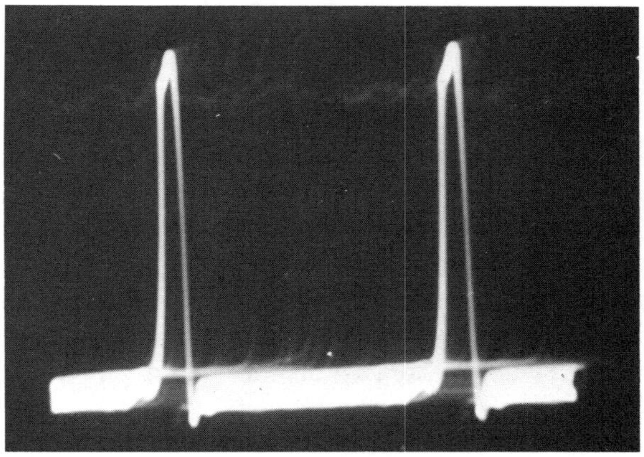

(2) SYNC PULSES.

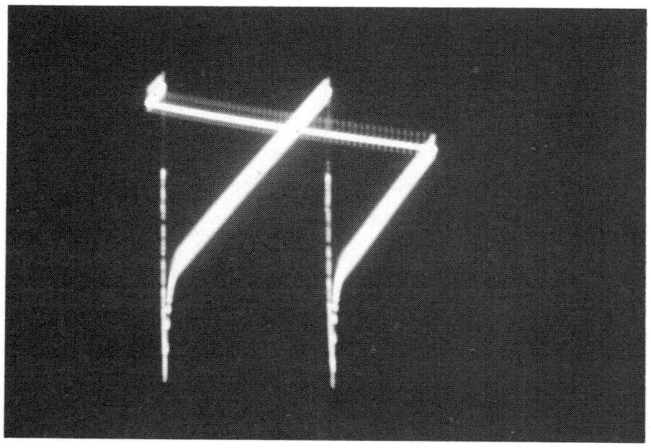

(3) VERTICAL-SWEEP SIGNAL.

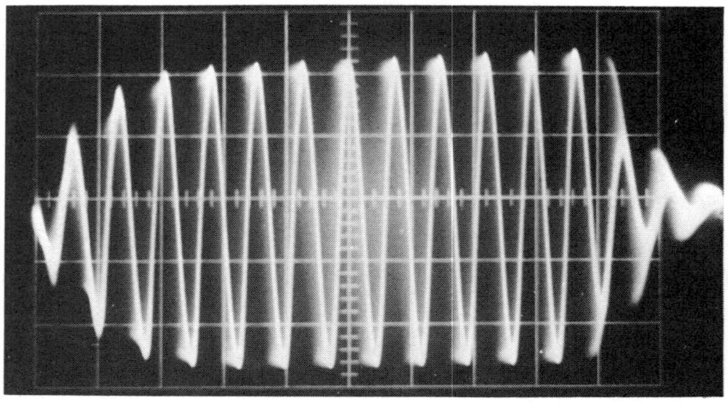

(4) COLOR-BURST WAVEFORM.

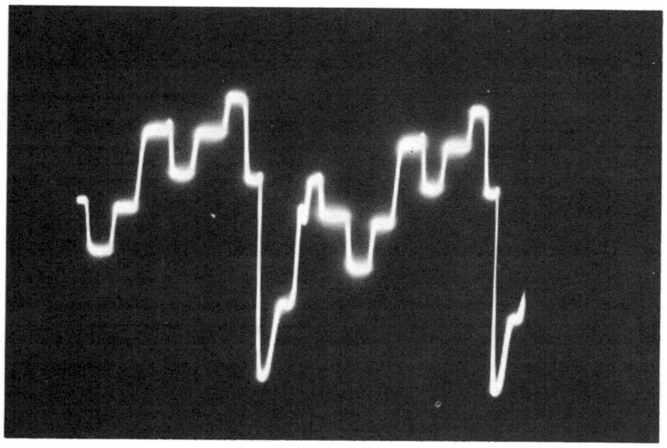

(5) COLOR-DEMODULATOR OUTPUT.

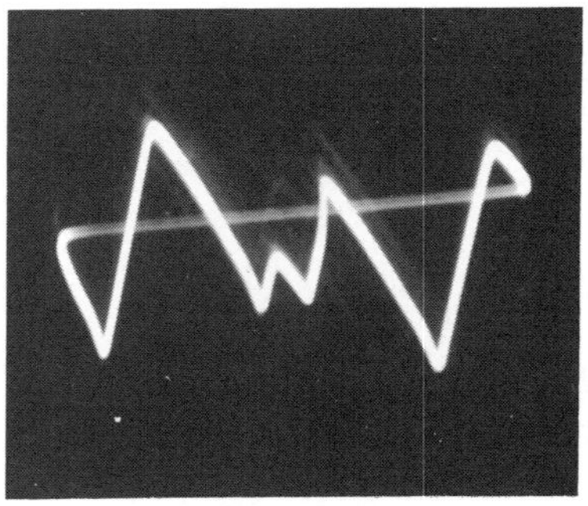

(6) POWER-SUPPLY RIPPLE.

(A) Photographs of waveforms.

Fig. 9-5. Various waveforms

9-5 Troubleshooting with a Scope

A scope is used very much like a voltmeter in troubleshooting. There is, however, one very important difference. The voltmeter reads only operating levels. It does not tell how well a particular stage or component is handling the signal. To illustrate the importance of this point, consider the

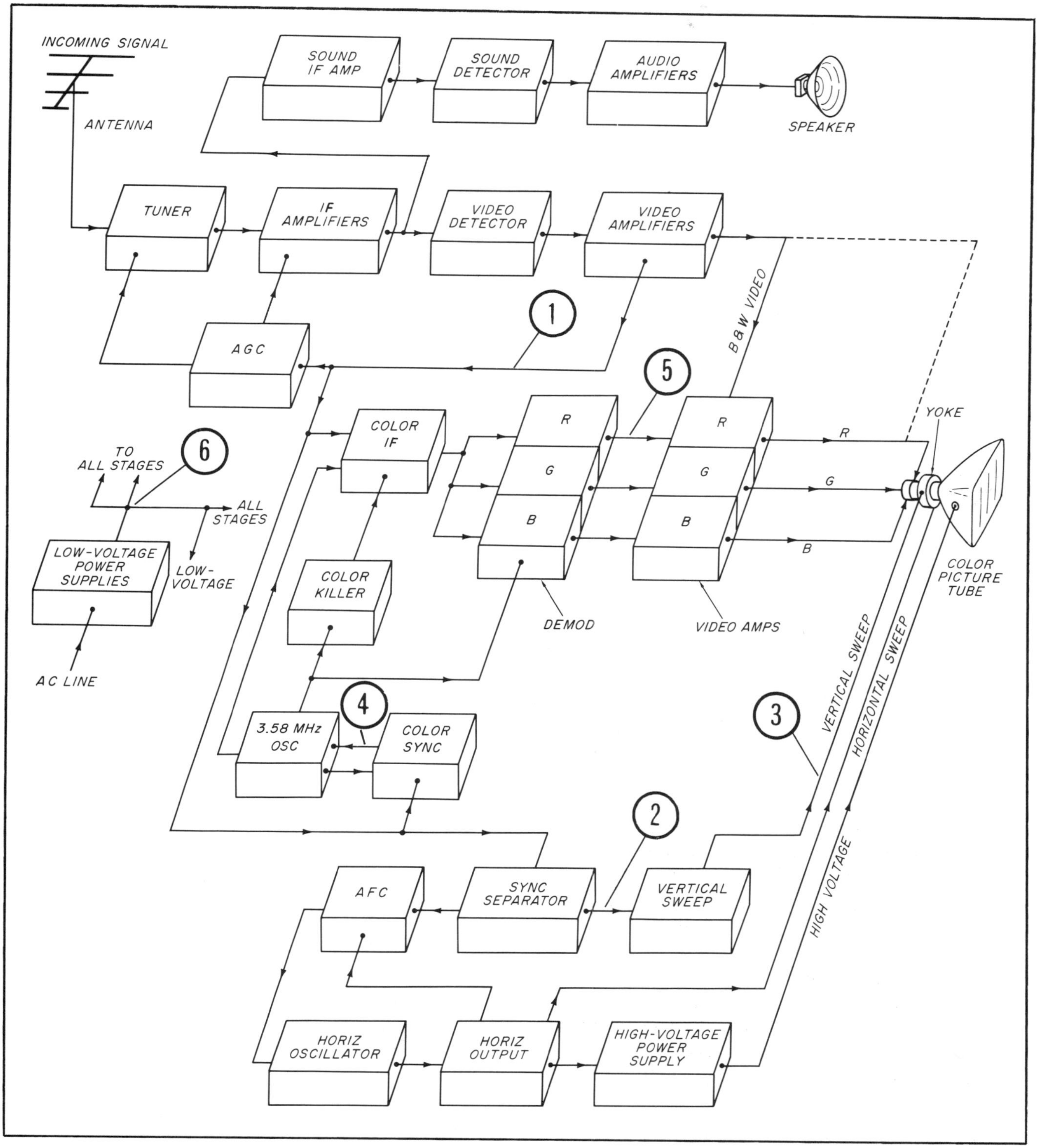

(*B*) *Waveform test points.*

found in television sets.

audio stages shown in Fig. 9-6. Suppose that the sound output level is very low and that the problem is caused by an open capacitor. The guilty capacitor is the 10-μF unit between the driver collector and the output base. This problem will not disturb the operating voltages of either transistor. So, if you rely only on voltage readings, you will not find the problem. However, suppose you take the oscilloscope and look at the waveform at the driver base—it seems OK. Then look at the waveform at the collector of the driver. It is bigger, so the stage is amplifying. Moving the scope leads to the base of the output transistor shows the signal *severely* reduced. The only thing likely to cause this is an open coupling capacitor.

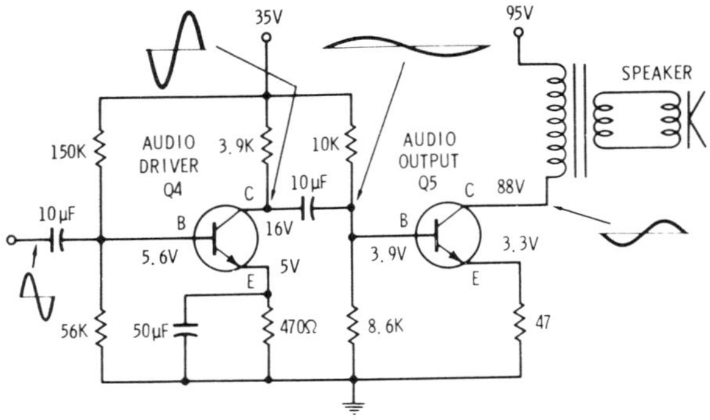

Fig. 9-6. An audio amplifier.

By observing the waveforms at the input and output of each stage, the operation of the stage and of the coupling networks between the stages can be checked. When this technique is combined with the use of a voltmeter to check operating voltages, you have a very powerful troubleshooting approach.

The normal waveforms at various test points are usually shown on the schematic diagram. With these waveforms, actual set operation can be compared against normal operation just as with voltage readings. You should keep in mind that there will be some variation from set to set, so don't expect your waveforms to be *exactly* like those shown on the diagram.

There are several set diagrams at the end of this book. Turn back to them now and examine the waveforms shown. Note the similarity between corresponding waveforms for two of the sets.

9-6 Voltage Measurement With an Oscilloscope

An ac voltmeter usually measures the *effective* value of a voltage. It does not follow the rapid changes in voltage waveform. Most voltmeters are calibrated for use only with dc and sinusoidal ac waveforms. They cannot be relied on for accurate measurement of the complex waveforms found in tv sets.

An oscilloscope "paints" a reasonably accurate picture of even very complex waveforms. If you know how many volts it takes to move the spot one inch vertically on the screen, you can measure the height of the waveform and figure its voltage. This is exactly what is done. Television waveforms should always be measured in terms of peak-to-peak volts. The peak-to-peak voltage is the voltage needed to move the oscilloscope spot from the very bottom of the waveform to the very top of it. For example, the peak-to-peak voltage of the waveform in Fig. 9-1 is 15 volts.

Finding out how many volts it takes to move the oscilloscope spot one inch is called *calibrating* the vertical deflection. The calibration procedure varies from one scope to another. Read the operation manual for your scope to find out how to calibrate it. You will find it to be a very helpful piece of information.

9-7 Review Questions

1. Explain in your own words what an oscilloscope is used for.
2. Explain why an ordinary voltmeter cannot do the same thing as an oscilloscope.
3. What is the peak-to-peak voltage of the waveform in Fig. 9-1?
4. Make a sketch showing how the waveform in Fig. 9-1 would look on an oscilloscope.
5. Explain how an oscilloscope crt works.
6. How is a scope tube different from a television picture tube?
7. What do the deflection plates in a scope tube do?
8. What are the basic sections found in an oscilloscope?
9. Which controls are normally found in the oscilloscope power-supply circuits?
10. Explain internal sync for an oscilloscope.
11. What does positive sync mean?
12. What does the vertical-sensitivity control do?
13. What is a horizontal-sweep generator for?
14. Explain how a scope is used in troubleshooting a tv set.
15. What kind of tv waveforms would you normally be interested in viewing?

16. What does the intensity control do to the spot on the scope tube?
17. Which control in a tv set does the same thing as the intensity control in an oscilloscope?
18. What do the scope positioning controls do?

CHAPTER **10**

Television Sync Problems

The television station sends out a series of synchronizing pulses that are used to lock the picture at the receiver in step with the picture at the station. Inside the tv receiver, these pulses are separated from the video signal by a section called the *sync separator*. Fig. 10-1 shows the location of this section in the set.

Inside the sync-separator block, there are three basic operations. Fig. 10-2 shows these operations. The actual sync-separator circuit removes the sync pulses from the video signal. The vertical integrator and the vertical-sync amplifier separate the vertical-sync pulses and send them to the vertical-sweep circuits. The horizontal differentiator and

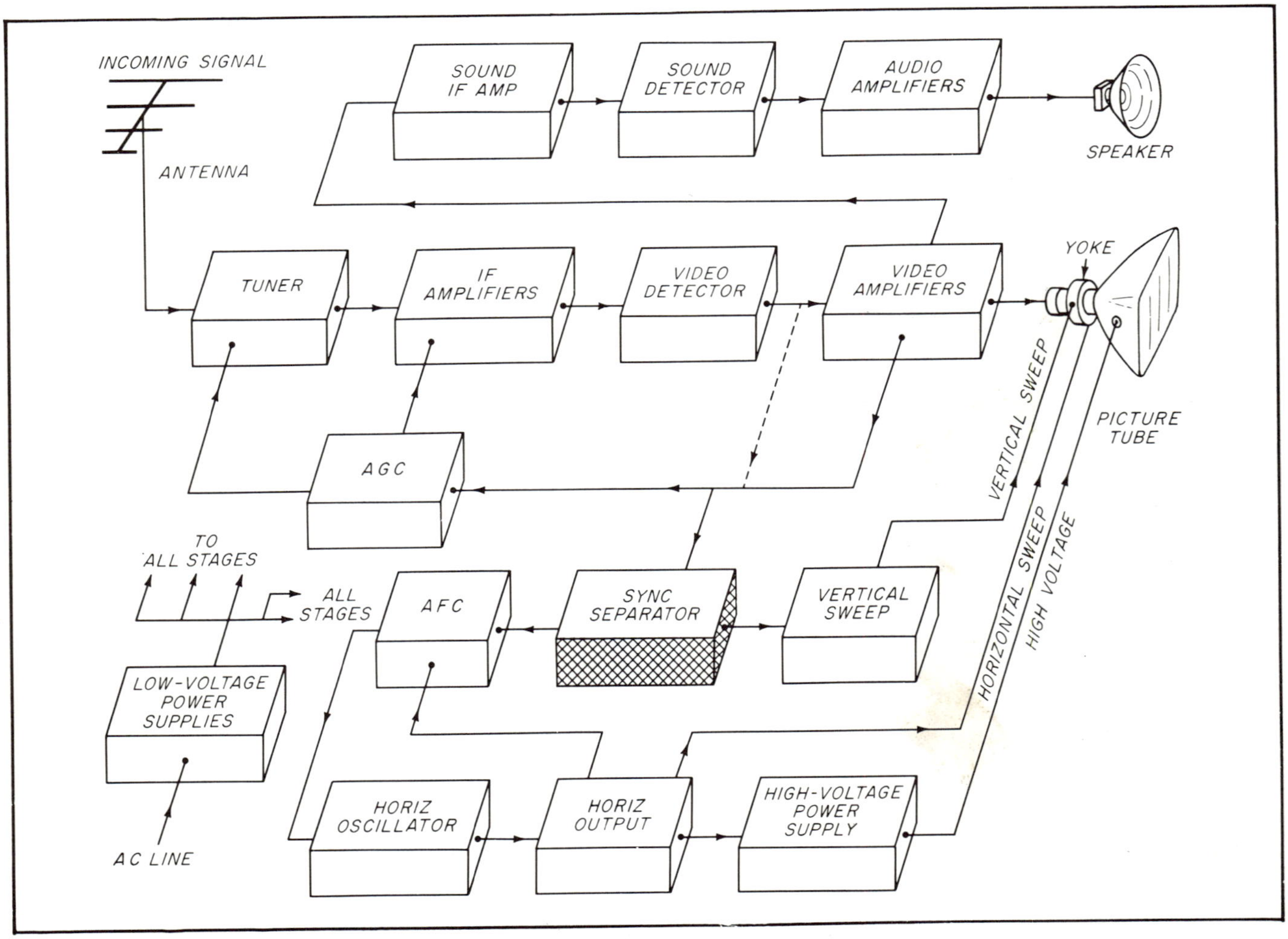

Fig. 10-1. Block diagram of a typical black-and-white receiver.

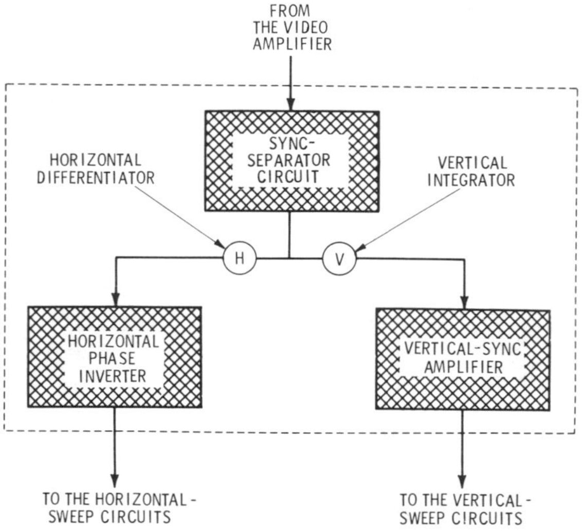

Fig. 10-2. The sync-separator system.

phase inverter separate the horizontal-sync pulses and send them to the horizontal-sweep circuits.

The horizontal- and vertical-sync pulses are used in the sweep circuits to lock the picture in with the station.

10-1 Loss of Vertical Sync Only

If the picture locks in horizontally but not vertically, then the failure must be in the vertical integrator or vertical-sync amplifier. Everyone has seen a picture like Fig. 10-3, which is out of vertical sync.

When there is a failure in the vertical side of the sync-separator system, the picture cannot be locked in with the vertical-hold control. By varying the vertical-hold control, it is possible to make the picture roll up or down, but not to lock it in. Notice that the important point is that the picture will roll both up and down but will not lock in.

Fig. 10-3. Loss of vertical sync.

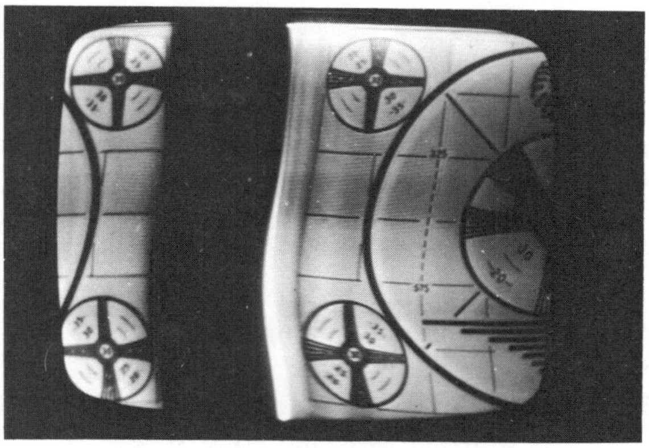

Fig. 10-4. Loss of horizontal sync.

As you will see later, failures in the vertical-sweep section can cause the picture to roll up or roll down, but not both.

10-2 Loss of Horizontal Sync Only

If the set will lock in vertically but not horizontally, then the problem must be in the horizontal differentiator, phase inverter, or afc circuit. Fig. 10-4 shows such a symptom.

Fig. 10-5. Severely out of horizontal sync.

If you vary the horizontal-hold control, you will be able to vary the horizontal sync all the way from severely torn out one way as seen in Fig. 10-5, through sync, to severely torn out the other way; but you will not be able to get it to hold. The important symptom is this ability to change the sync from severely torn out one way, through sync, to severely torn out the other way. You will see later that there can be failures in the horizontal sweep section that will cause the picture to tear one way or the other, but you will not be able to vary it from one to the other.

10-3 Loss of Vertical and Horizontal Sync

If there is a failure in the sync-separator circuit itself, then both horizontal and vertical sync will be lost at the same time. Fig. 10-6 shows this symptom. This symptom is just the combination of the previous two. If the vertical-hold control is varied, the picture will roll up and down while at the same time moving sideways. It will not lock in. If the horizontal-hold control is varied, the picture can be made to tear, first one way and then the other, and at the same time it will be rolling vertically. If the two hold controls are set very carefully, the picture can be made to look normal for a few seconds, but after a very short time it will start drifting off again.

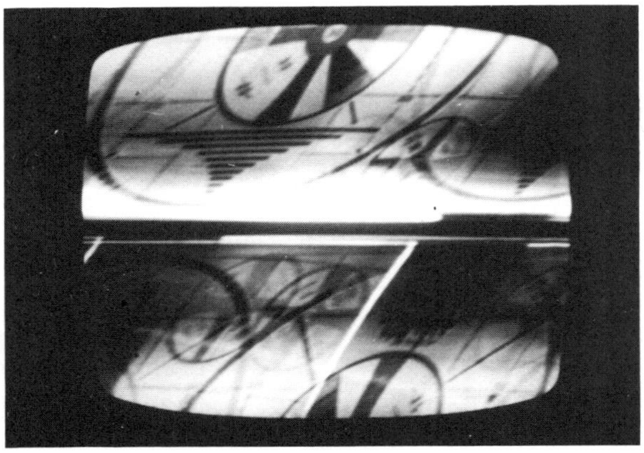

Fig. 10-6. Loss of both horizontal and vertical sync.

The thought may have occurred to you that the same symptoms would appear if the sync pulses were lost before they got to the sync separator stage. This is, of course, correct, so what is needed is a way to check to see if the sync pulses are getting through. An oscilloscope can be used to look at the sync. Fig. 10-7 shows the video signal at the input of the sync separator. The bright horizontal lines at the top of the waveform are the horizontal-sync pulses. The bright vertical lines at the top of the waveform are the vertical-sync pulses. The wavy pattern in the bottom three-fourths of the waveform is the actual video information. A look at the sync pulses after they have been separated from the video will show a waveform like Fig. 10-8.

While using the oscilloscope to trace sync pulses is very effective when the set is out of its cabinet, you still need a way to check the pulses without removing the set. Possibly you have noticed the black bar that separates the pictures when the set

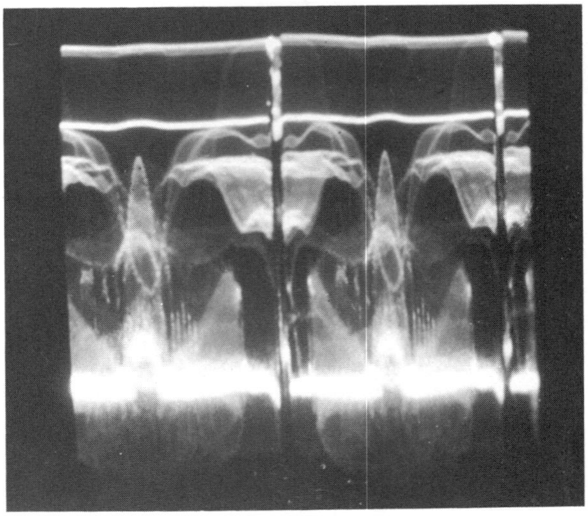

Fig. 10-7. Video at the sync-separator input.

is out of sync vertically. That bar is called the *vertical blanking* bar. In most sets, the screen is blanked out during the time between picture fields. This allows the vertical sweep to retrace (go back to the top of the screen) and get ready to start the next field. The vertical-sync pulses occur during this retrace time, and since the sync pulses go through the video amplifiers to the picture tube, the vertical-sync pulse is on the screen inside the blanking bar. It usually cannot be seen because the blanking bar is so black. On many sets (but not all), you can let the picture roll slowly and turn the contrast down to make the sync pulse visible. Fig. 10-9 shows what it looks like. This sync-pulse pattern is called the *vertical hammerhead* pattern by many technicians. If it is quite black compared to the blanking bar, then the sync pulse is strong. On some sets you won't be able to see the hammerhead because the contrast control will not make the blanking bar light enough. Keep in mind that

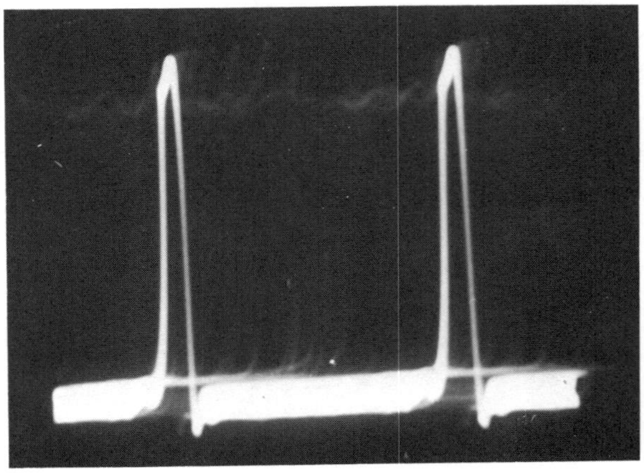

Fig. 10-8. The separated sync pulses.

this hammerhead pattern comes from the station. It will only be seen on channels where there is a program being telecast.

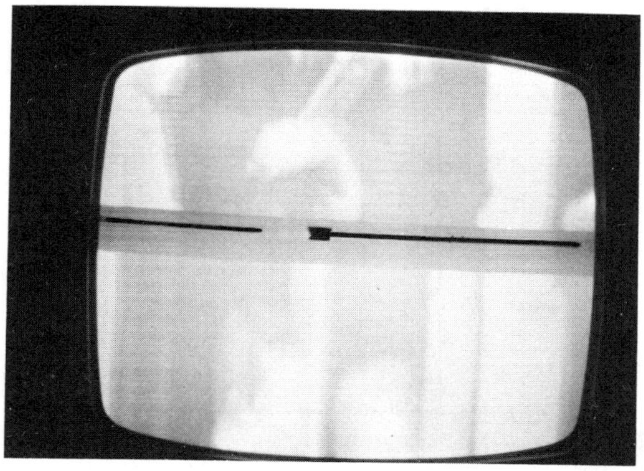

Fig. 10-9. Vertical-sync pulses in the blanking bar.

10-4 Sync-Separator Circuitry

The circuits used for sync separation are different in almost every set. However, they all work about the same. Fig. 10-10 shows a typical circuit. The first sync-separator stage is biased well into cutoff so that only the very tips of the sync pulses are able to turn it on. This action allows the sync-pulse tips to be amplified while preventing the video from passing.

Many sets have only one sync-separator stage. In such a case, the sync pulses from the single-stage sync separator are coupled immediately to the horizontal- and vertical-sync stages.

In Fig. 10-10, there are two sync-separator stages shown. The output of the first stage is coupled into the second stage. The operation of the second stage is the same as that of the first. It again allows only the sync-pulse tips to be amplified and removes any video that may have leaked through the first stage.

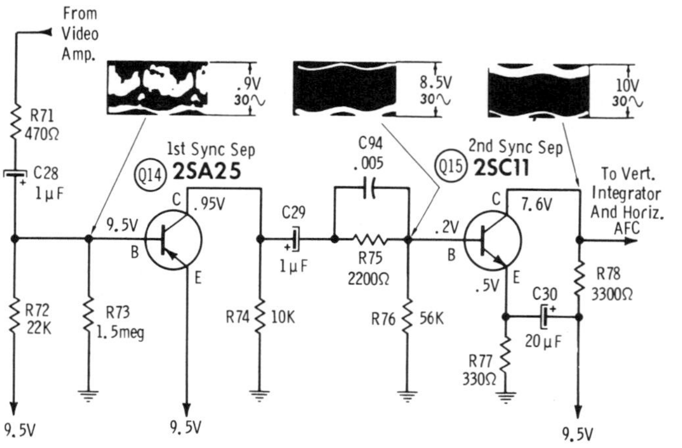

Fig. 10-10. A sync-separator circuit.

From the sync separator, the sync signal is sent to the vertical- and horizontal-sync stages. Fig. 10-11 shows an example of these circuits. In this example, there is an additional stage of sync amplification in Q201. Many sets do not have this additional sync amplifier. If the sync amp is defective, both horizontal and vertical sync will be lost. From the collector of the sync amplifier, the pulses are applied to both the vertical- and horizontal-sync circuits.

The vertical-integrator circuit consists of R203 and C202. This simple circuit is a low-pass filter; that is, it will allow low-frequency signals to pass but blocks high-frequency signals. The vertical-sync pulses are a 60-Hz signal, but the horizontal ones are a 15,750-Hz signal. The vertical integrator allows the low-frequency vertical-sync signal to pass through to the vertical-sync amplifier. The higher-frequency horizontal-sync signals, on the other hand, are blocked. They cannot get through to the vertical-sync amplifier. The vertical-sync pulses are amplified by Q202, which is the vertical-sync amplifier. Then they are sent on to the vertical-sweep circuit through transformer T201.

The sync amplifier (Q201) is connected to the horizontal differentiator composed of C212 and R225. This circuit is a high-pass filter. It will only pass the high-frequency horizontal-sync pulses. It blocks the low-frequency vertical pulses. The horizontal-sync pulses go through the horizontal phase inverter (Q205) and are fed to several points in the horizontal-sweep circuit through C213 and C214.

While exact circuit details vary from one set to another, most sets have sync-separator sections that work almost the same as this one. There are circuit diagrams of several sets at the back of the book. You should examine the sync-separator system in each of them to become familiar with some of the possible circuit variations.

10-5 Troubleshooting the Sync Section

Once a trouble has been isolated to the sync section, troubleshooting presents no particular problem. Circuit diagrams normally show both dc operating voltages and waveforms.

All that is necessary is to measure the operating voltages and compare them to the values given on the diagram. If this does not reveal the source of

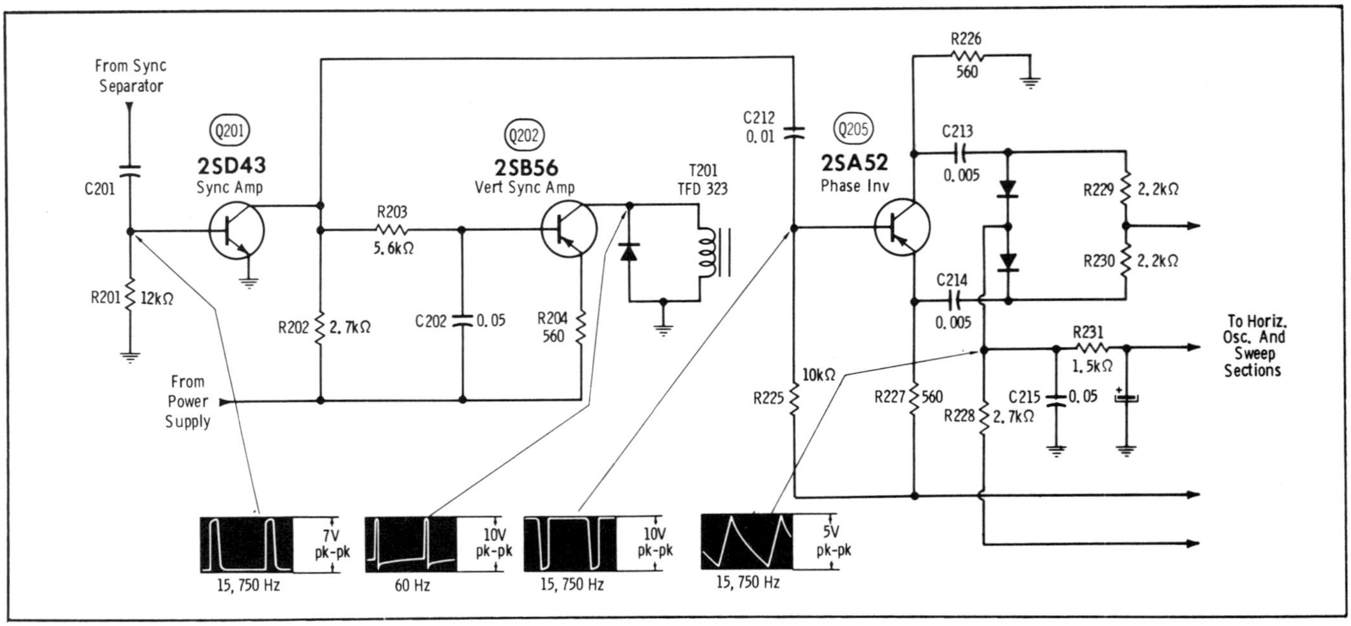

Fig. 10-11. Vertical- and horizontal-sync circuits.

the trouble, use the oscilloscope to view the waveforms, and compare them to the ones shown on the diagram.

As an example, suppose that the set containing the circuit of Fig. 10-11 will not hold horizontal sync. Varying the horizontal-hold control reveals that the sync can be shifted from torn one way to torn the other way. The horizontal differentiator or the phase inverter should be suspected. Since operating voltages are not shown in Fig. 10-11, you would go directly to viewing the waveforms. The waveform at the base of Q205 agrees with the one shown on the diagram. It, therefore, should be concluded that the problem must be in the phase inverter. When transistor Q205 is checked, it is found to be open-circuited. Replacing Q205 cures this problem.

As another example, suppose that the set containing the circuit of Fig. 10-10 has no vertical or horizontal sync. Varying the hold controls suggests a bad sync-separator circuit. The hammerhead pattern is checked, and it is normal.

A check with a vom (voltmeter) shows the operating voltages in Fig. 10-10 all to be normal. Look at the video at the base of Q14—it is normal. The waveform at the base of Q15 is found to be very weak (only about 1 volt peak-to-peak). Now you are closing in on the problem. Look at the waveform at the collector of Q14. There, the sync pulses are almost 10 volts peak-to-peak. The only thing that could cause the loss of the pulses between these two points is either C29 or R75. Check the capacitor first by replacing it with a new one—yes, that fixed the set.

By use of these troubleshooting methods, problems that affect the sync-separator section can usually be found with little trouble.

10-6 Review Questions

1. What happens to a set if the vertical-sync amplifier fails? Describe the symptoms.
2. Transistor Q201 in Fig. 10-11 is bad. What symptoms would you expect to see on the set?
3. What is a hammerhead pattern?
4. The set containing the circuit of Fig. 10-11 has no horizontal sync. The waveform at the base of Q205 is only about one-half volt peak-to-peak. The waveform at the base of Q201 is normal and so is the vertical sync. What could be the problem?
5. Explain in your own words what the vertical-blanking bar is and how you can see it on the screen.
6. What is the function of the sync separator in a television set?
7. The collector-to-emitter voltage of Q202 in Fig. 10-11 is zero. What would the set symptoms be? What do you think the trouble is?
8. How do you adjust the set to see the hammerhead pattern?
9. What would be the symptoms if Q205 in Fig. 10-11 failed?

10. The waveform at the base of Q15 in Fig. 10-10 is normal. At the collector of Q15 there is no viewable waveform. What would you check next and why?
11. What would be the symptoms of the set in Question 10?
12. What would you do first if a set had the symptoms shown in Fig. 10-5?
13. The waveform at the base of Q15 in Fig. 10-10 is normal, but there is no sync as far as the picture is concerned. What could cause the problem? How would you check your answer?
14. What would you do to diagnose the problem shown in Fig. 10-6? List everything you would do before removing the back cover of the set.
15. What does the vertical integrator do in a television set?
16. Which components in Fig. 10-11 are part of the vertical integrator?
17. Explain which parts of Fig. 10-7 are:

    A. The video information.
    B. The vertical-sync pulses.
    C. The horizontal-sync pulses.

18. Which stages in the sync-separator system could cause the symptoms shown in Fig. 10-4?

CHAPTER **11**

Picture Sweep Failure

The basic function of a sweep section is to move the spot of light across the screen of the picture tube. In a tv set, there are two separate sweep sections.

11-1 The Sweep Sections

The vertical-sweep section indicated in Fig. 11-1 sweeps the spot from the top of the screen to the bottom and then very quickly retraces (returns the spot) to the top. The vertical-sweep section does this 60 times every second, and, as a result, this section is said to operate at a frequency of 60 Hz.

The horizontal-sweep section, shown in Fig. 11-2, sweeps the spot from the left side of the screen (as seen by the tv viewer) to the right side. Then it quickly retraces to the left side. The

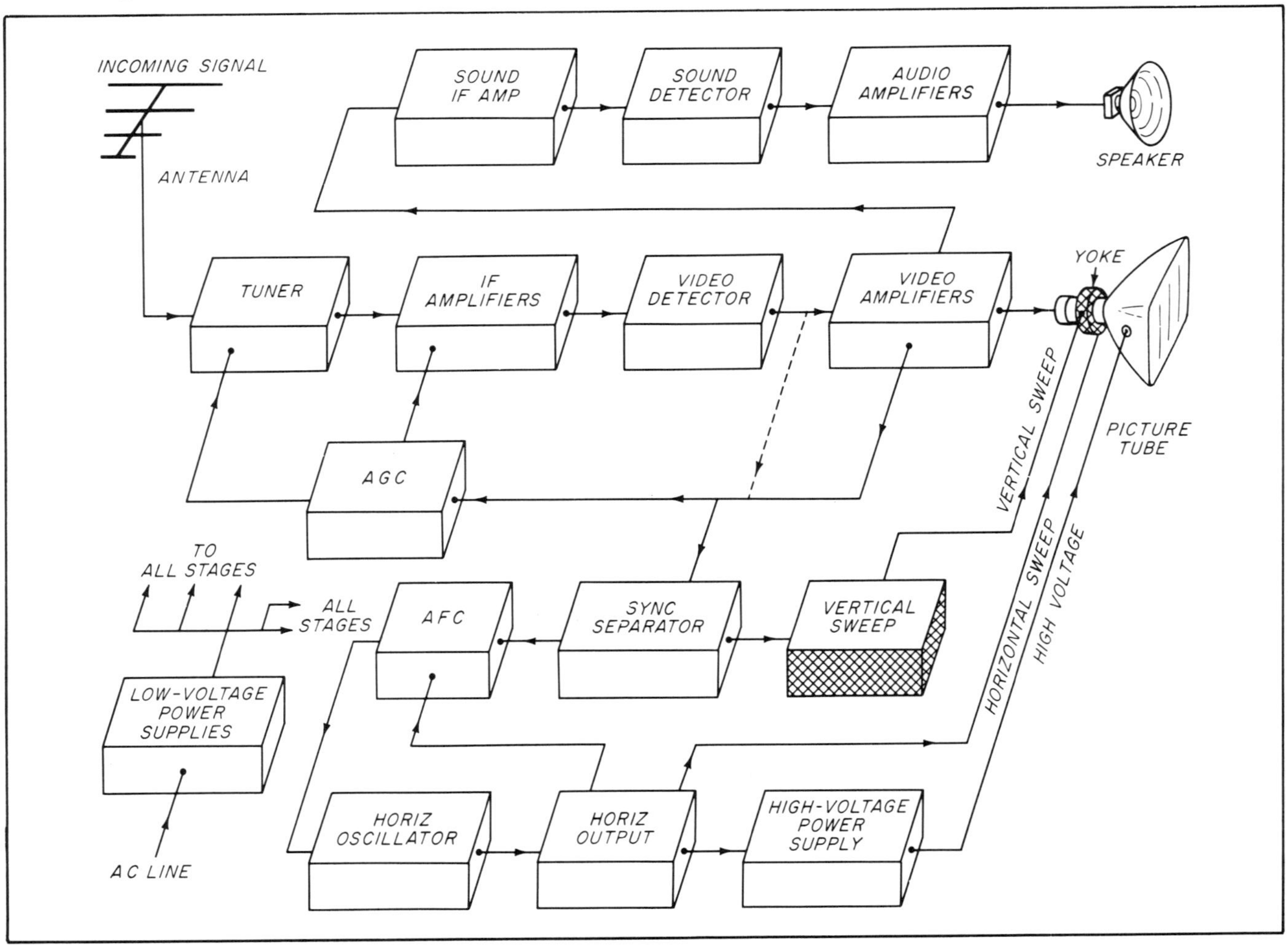

Fig. 11-1. Typical black-and-white receiver with vertical-sweep section indicated.

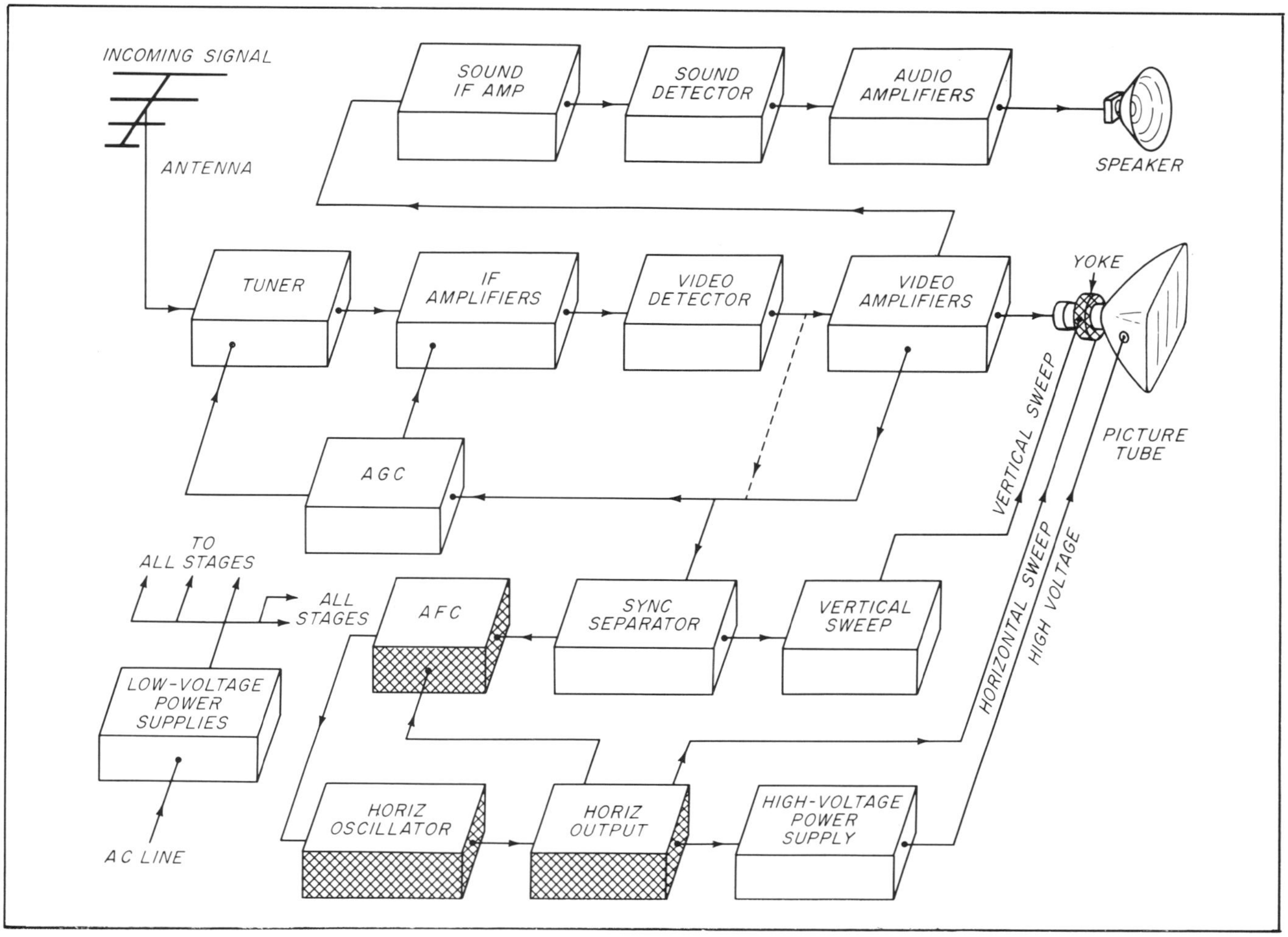

Fig. 11-2. Block diagram of receiver with horizontal-sweep section indicated.

horizontal-sweep section goes through this cycle 15,750 times a second; thus it is said to operate at 15.75 kHz.

11-2 Vertical-Sweep Failures

Inside the vertical-sweep section you may find either two or three stages. Fig. 11-3 shows the block diagrams of both of these types of vertical-sweep sections.

With either type of vertical-sweep section, three basic types of vertical-sweep failures can occur. These basic types of failures are:

1. Loss of vertical sync
2. Loss of picture height
3. Poor vertical linearity

Consider each one of these types of failure separately.

11-3 Loss of Vertical Sync

Suppose that a set displays the symptoms shown in Fig. 11-4, and when you vary the vertical-hold control there is little or no change in the picture sync. The important thing to notice here is that the vertical-hold control does not change the sync enough to make the picture roll *both* up and down.

This problem is almost surely in the vertical-oscillator stage. The oscillator is not operating at the correct frequency (60 Hz). If the oscillator frequency is too high, the picture will roll downward; if the frequency is too low, the picture will roll upward.

Fig. 11-5 shows the circuit diagram of a typical two-stage vertical section. Notice that feedback for the oscillator comes through transformer T2. The frequency of operation is determined by the components in the base circuit of the vertical-oscillator transistor (Q17). These components include the transformer winding, R4, R89, R90, and

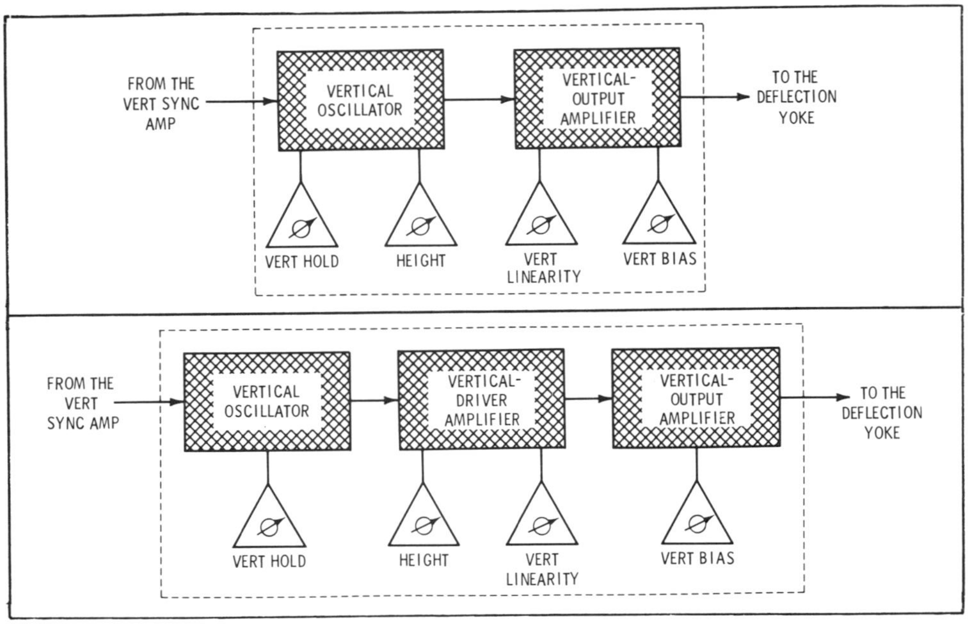

Fig. 11-3. Typical vertical-sweep sections.

C22. If any of these parts changes value enough, it can cause the out-of-sync symptoms. The vertical-hold control (R4) is perhaps the most frequent offender.

Some recent sets use a *countdown* system to provide improved vertical sync. In this system, the horizontal pulses are electronically counted, and a vertical pulse is generated for each field. These pulses are compared to the separated vertical-sync pulses to ensure correct synchronization.

11-4 Loss of Picture Height

Poor picture height is caused by weak output from the oscillator or reduced gain in the vertical output or driver stages. The symptom may be only slightly reduced height as shown in Fig. 11-6.

Fig. 11-4. Loss of vertical sync.

Notice that the height control in the two-stage section (Fig. 11-5) simply controls the bias on the oscillator transistor. You will recall from Chapter 8 (automatic gain control) that the gain, and hence the output signal level, of a transistor can be controlled with the bias.

It is worth noting that both of the vertical sections shown here use transformer feedback in the oscillator stage. This is not always the case. There are several ways to develop the sweep signal without employing transformer feedback, and these circuits are used in some sets.

Another way of controlling height is shown in Fig. 11-7. Here the vertical-size (height) control simply determines the amount of the oscillator signal that is coupled to the driver amplifier. Weak transistors or tubes are probably the most frequent cause of reduced picture height. However, any other component failure that would cause a signal-level reduction could cause this problem.

If any of the stages in the vertical-sweep section fail completely, resulting in a total loss of sweep signal, the result will be the symptom shown in Fig. 11-8. Here you see a complete loss of height, indicating no vertical sweep at all.

11-5 Poor Vertical Linearity

Poor vertical linearity like that seen in Fig. 11-9 means that the vertical-sweep section has a different sweep rate in different parts of the picture. In Fig. 11-9A, the vertical-sweep section is

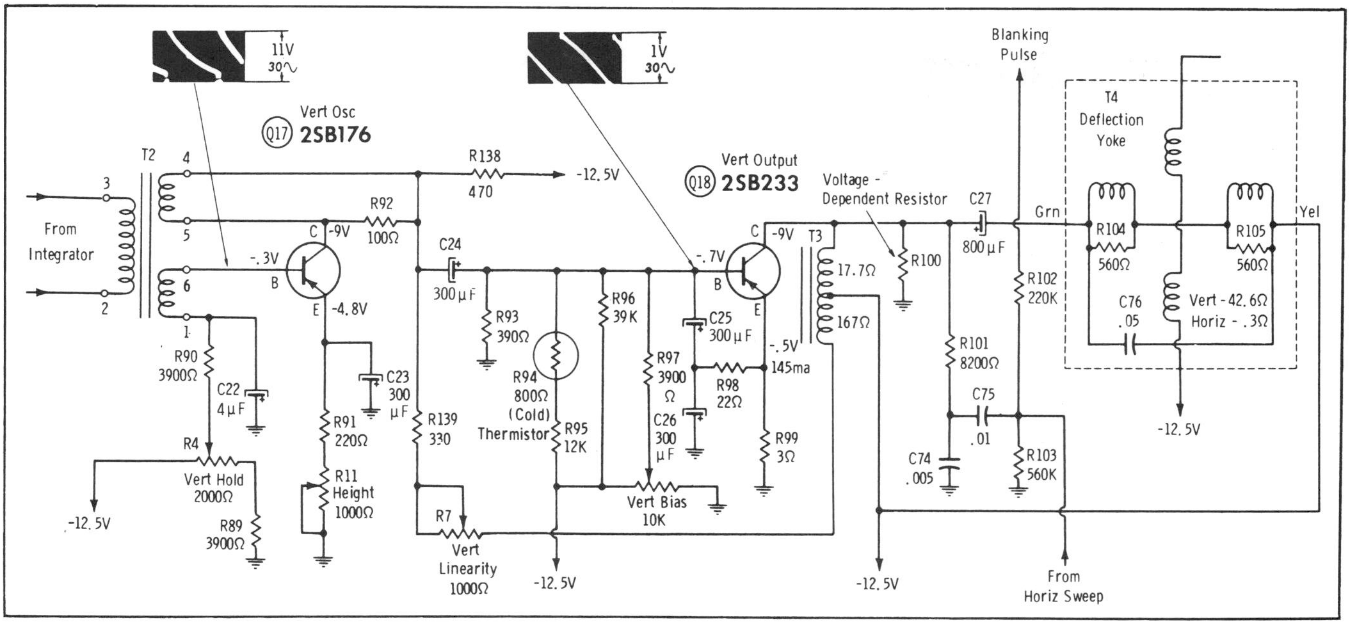

Fig. 11-5. A two-stage vertical-sweep section.

sweeping faster at the top than it is at the bottom. In Fig. 11-9B, the sweep rate is faster at the bottom than it is at the top.

Inside the sweep circuits, this vertical nonlinearity shows up as an incorrectly shaped vertical-sweep signal. Fig. 11-10 shows a properly shaped sweep signal and an improperly shaped one. The nonlinearity shows up as a curved section in the vertical-sweep waveform. Notice that the sloping top of the properly shaped waveform is straight. The other one is curved near the top,

Fig. 11-6. Reduced height.

indicating the presence of nonlinearity at the bottom of the screen. (The waveforms in Fig. 11-10 also contain undesired horizontal pulses, which will affect interlace.)

There are two controls in the vertical circuit that affect linearity. They are:

1. The *vertical linearity control.* While this control influences the linearity over the whole screen, it has its most obvious effect at the top of the picture.
2. The *height* control. This adjustment sets the picture size vertically. It tends to have the most effect at the bottom of the picture.

In most sets, both controls interact with each other. Consequently, you should adjust them alternately for the best overall picture.

The most common causes of nonlinearity are defective capacitors or transistors in the driver or output stages.

Another failure that can occur in the vertical-sweep circuit is a short circuit inside the deflection yoke. This failure, while occurring only rarely, causes the symptom shown in Fig. 11-11. This symptom is very easy to spot. Notice that the picture is shaped like a *keystone;* hence, we call it a *keystoned picture.*

11-6 Vacuum-Tube Vertical-Sweep Sections

Vertical-sweep circuits employing vacuum tubes are quite similar to those using transistors. Fig. 11-12 shows a two-stage vertical section that is very much like Fig. 11-5.

Vacuum tubes are usually able to handle more power than transistors. Therefore, a three-stage vacuum-tube circuit is not usually required. In fact, it is possible to have a section with only one

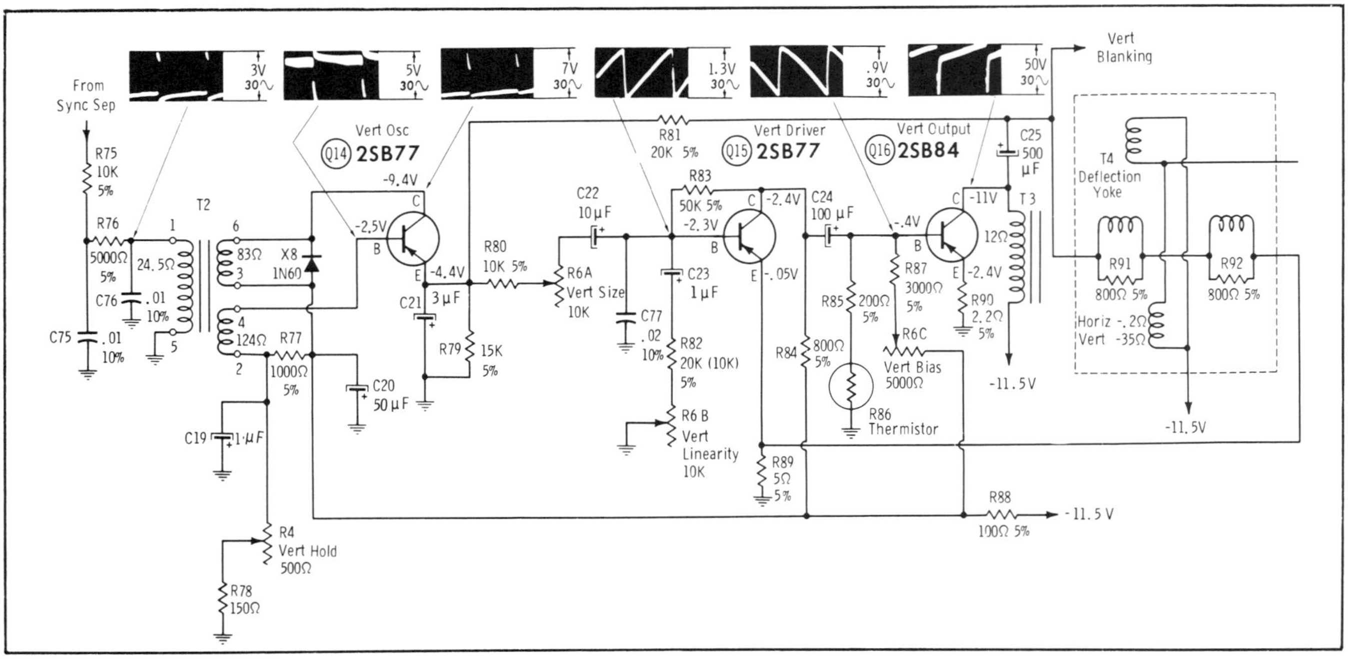

Fig. 11-7. A three-stage vertical-sweep section.

vacuum tube. Fig. 11-13 shows such a single-tube circuit. The multivibrator circuit shown in Fig. 11-14 is also frequently used.

11-7 Horizontal-Sweep Failures

The horizontal-sweep circuits are not entirely different from the vertical-sweep circuits. There are, however, a few very important differences. Fig. 11-15 shows a block diagram of the horizontal-sweep system.

Notice that the horizontal-sweep system has an automatic frequency control (afc) circuit, but the vertical circuit did not. This circuit compares the horizontal-output signal with the sync pulses. If there is any frequency difference, the afc circuit corrects the horizontal-oscillator frequency.

Fig. 11-8. No vertical sweep.

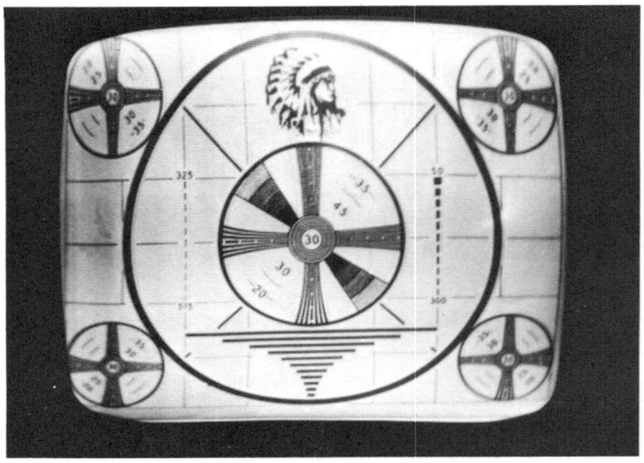

(A) Top expanded, bottom compressed.

(B) Bottom expanded, top compressed.

Fig. 11-9. Poor vertical linearity.

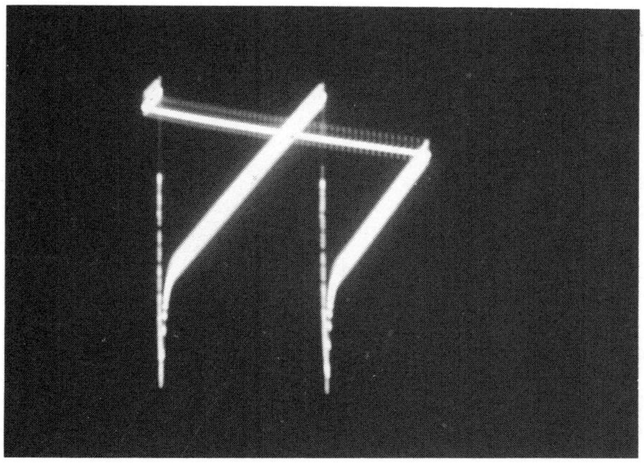

(*A*) *Proper shape.*

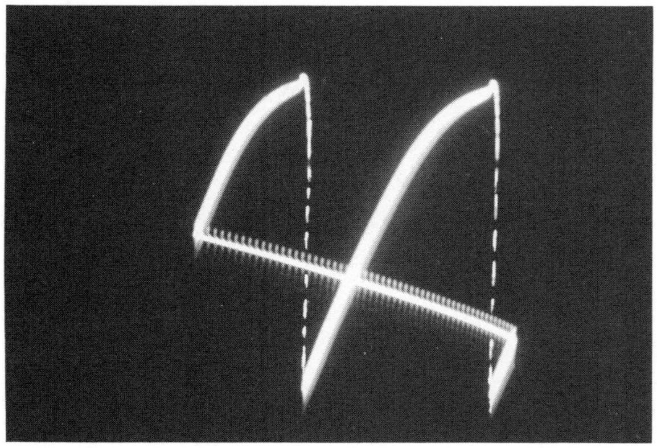

(*B*) *Nonlinearity present.*

Fig. 11-10. Vertical waveforms with and without nonlinearity.

Another major difference between the horizontal- and vertical-sweep circuits is that the horizontal-output stage supplies the input to the high-voltage power supply. This fact makes horizontal-sweep problems somewhat more complicated than those found in vertical sections.

A third important difference between the two sweep sections is that the horizontal section operates at a much higher frequency. The horizontal-sweep system moves the light spot across the screen 15,750 times a second, while the vertical-sweep section moves it up and down only 60 times a second. Moreover, the screen is wider than it is tall, so the horizontal section must move the spot farther than the vertical section does every cycle. The overall effect is that the horizontal section must handle much more power than does the vertical-sweep system. Higher power means higher energy levels and probably more frequent breakdowns.

The types of symptoms that you are likely to see with the horizontal-sweep section are more or

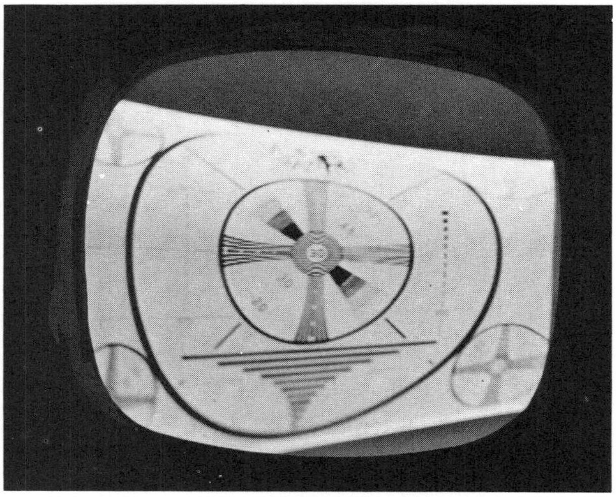

Fig. 11-11. Keystoned picture.

less similar to those discussed for the vertical system. The most common ones are:

1. Loss of horizontal sync
2. Loss of picture width
3. Poor horizontal linearity
4. Loss of raster

Let us consider each of these horizontal-sweep problems one at a time.

11-8 Loss of Horizontal Sync

The horizontal oscillator must operate at 15,750 Hz if the picture is to be in sync horizontally. When the picture is out of sync horizontally, *and the horizontal-hold control has little or no effect*, the oscillator should be suspected. Fig. 11-16 shows the usual symptoms.

However, since the oscillator frequency is determined by the afc circuit, the trouble could also be in that stage. Fig. 11-17 shows a typical afc-oscillator circuit. Another afc method is shown in Fig. 11-18. The afc circuit is composed of diodes X14 and X15 and their related parts.

The out-of-sync condition might be quite mild, as shown in Fig. 11-19, or more severe as in Fig. 11-16. The mild horizontal *pulling* is probably harder to repair simply because it is mild (everything is *almost* OK). If the horizontal-oscillator frequency is too far away from 15,750 Hz, there will be no raster at all on the screen. This problem will be discussed in more detail in Chapter 12.

The most common causes of loss of horizontal sync in the afc-oscillator section are defective afc diodes or a bad capacitor in the afc or oscillator circuit. Occasionally, the horizontal-oscillator coil will cause trouble, and, in rare cases, hori-

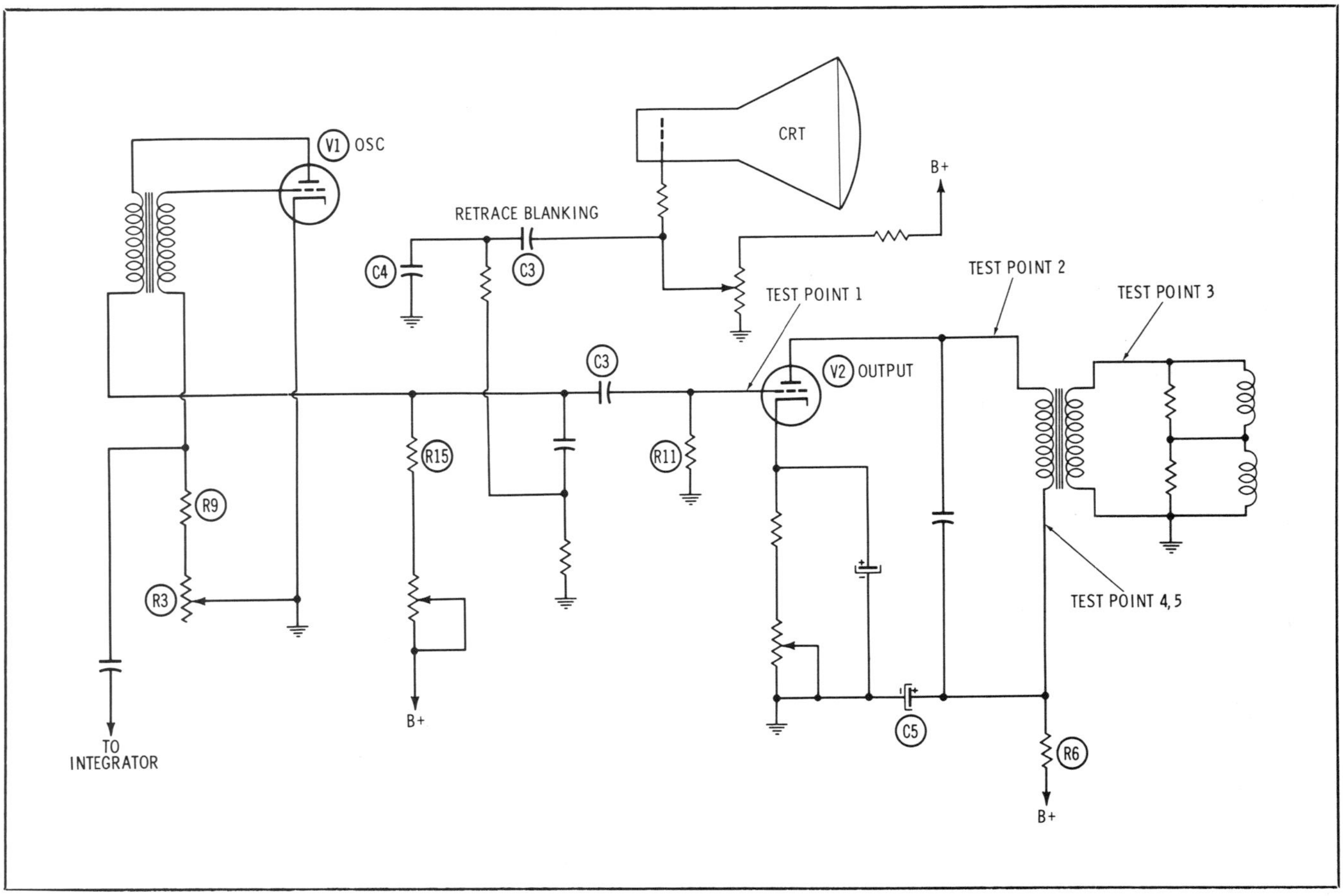

Fig. 11-12. A vacuum-tube two-stage vertical-sweep section.

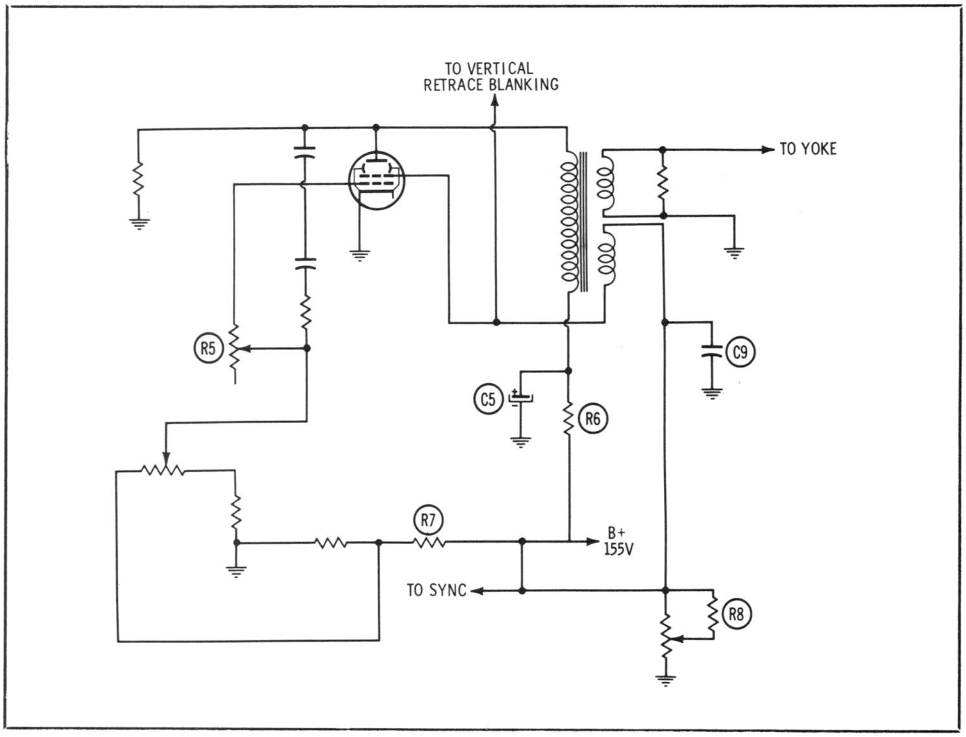

Fig. 11-13. Single-tube vertical-sweep section.

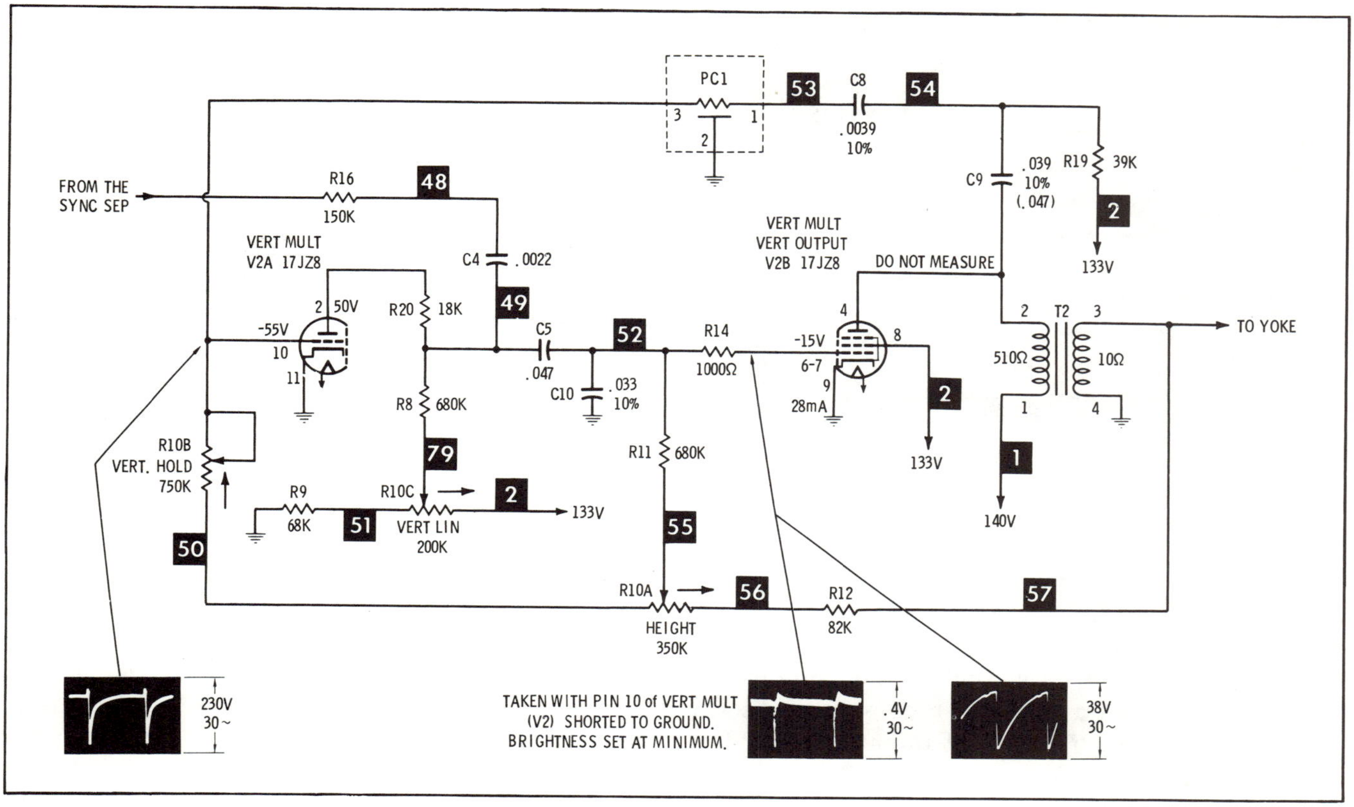

Fig. 11-14. Multivibrator vertical section.

zontal-sync trouble can be caused by a failure in the horizontal-output stage.

11-9 Loss of Picture Width

Fig. 11-20 shows a picture that is pulled in from both sides. Loss of width can be caused by a bad transistor in the horizontal-output circuit shown in Fig. 11-21 or by any other problem that reduces the horizontal-sweep signal. A weak horizontal-driver transistor or oscillator transistor, for example, can cause this problem. So can any other failure that tends to reduce the gain of one of the stages in the horizontal-sweep section.

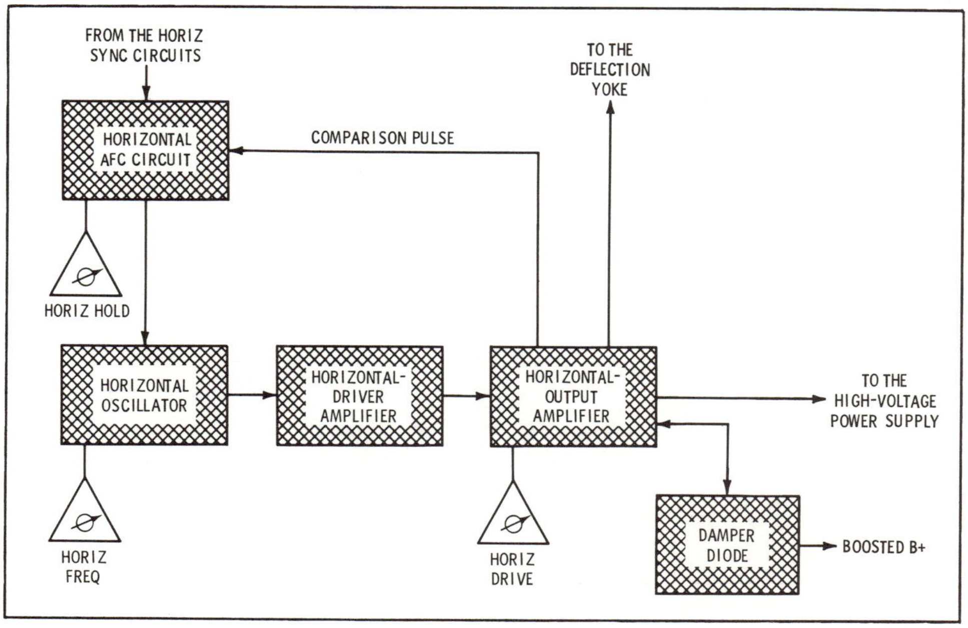

Fig. 11-15. Block diagram of horizontal-sweep section.

Fig. 11-16. Out of horizontal sync.

The horizontal-output transformer and deflection yoke are tuned circuits. They are tuned to about 50 kHz, or roughly 3 times the horizontal-sweep frequency. This high-frequency tuning is necessary to allow the light spot to be retraced rapidly enough. However, tuning the horizontal transformer causes a problem because it is then inclined to oscillate. To prevent these undesirable oscillations from interfering with the picture after retrace, a *damping diode* is connected across the output transformer.

In many sets, the action of this damping diode is used to produce a medium high voltage called the *boosted B+*, which in turn is used to operate various circuits that require high voltages. In some sets, the boosted B+ is used to operate both the horizontal- and vertical-output stages. In yet other sets, a separate boost diode is used to produce the boosted B+ voltage.

A damper-diode failure can cause a complete loss of raster, loss of width, or loss of width and height, depending on the type of circuit. Always check the damper when a horizontal-output failure is suspected.

Some sets use a horizontal-output stage that does not employ bipolar transistors at all. Fig. 11-22 shows such a circuit. In this circuit, a diode (D1) and a silicon controlled rectifier (SCR1) are used to produce the horizontal trace. Another pair (D2 and SCR2) are used to control the retrace.

Troubleshooting such a circuit is the same as for a transistor circuit. That is, once the trouble is isolated to the output circuit, each component is checked individually. You will recall that if a stage fails in the vertical-sweep section and completely cuts off the signal, the result is no vertical sweep (see Fig. 11-8). If a stage fails in the horizontal-sweep section, the result is no raster.

It is possible to have a symptom like the one shown in Fig. 11-23. This symptom can be caused

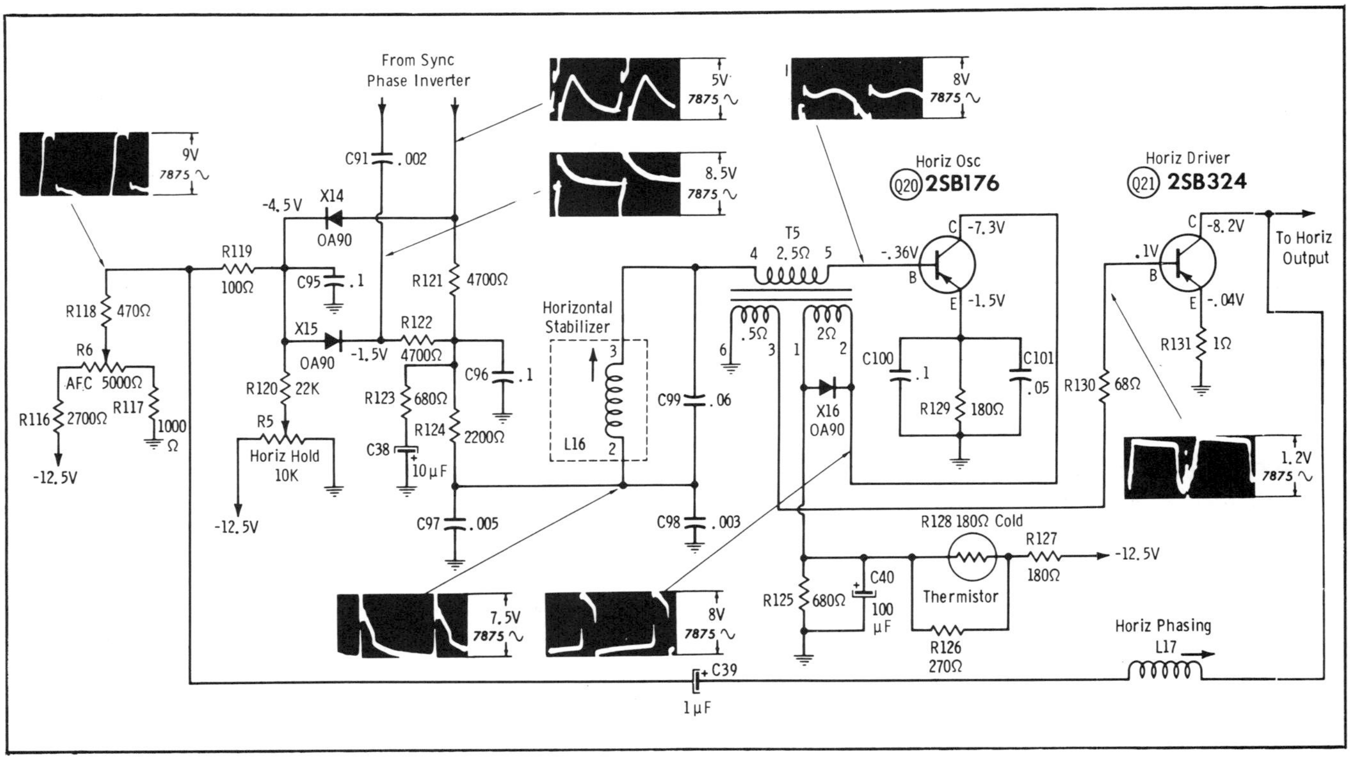

Fig. 11-17. An afc, oscillator, and driver section.

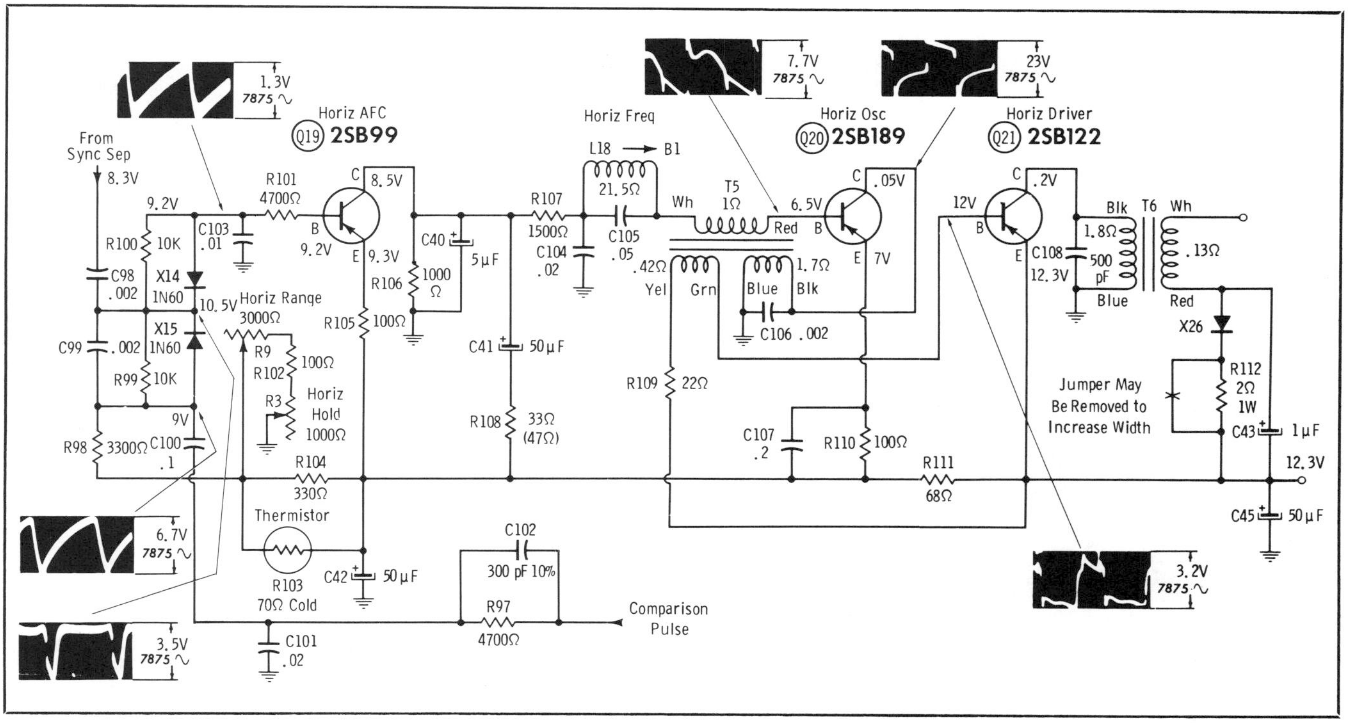

Fig. 11-18. An unbalanced transistor afc circuit.

only by an open circuit in the horizontal coils of the deflection yoke. If there is a short in the yoke (or sometimes in an output transformer), it will cause *keystoning*. The keystoning may be horizontal, or it may be vertical as shown in Fig. 11-24. However, these last two symptoms (Figs 11-23 and 11-24) are rather rare.

11-10 Poor Horizontal Linearity

Nonlinearity in the horizontal sweep (Fig. 11-25) is caused by improperly shaped horizontal-sweep pulses. The component at fault is always in the horizontal-output stage and is most often a

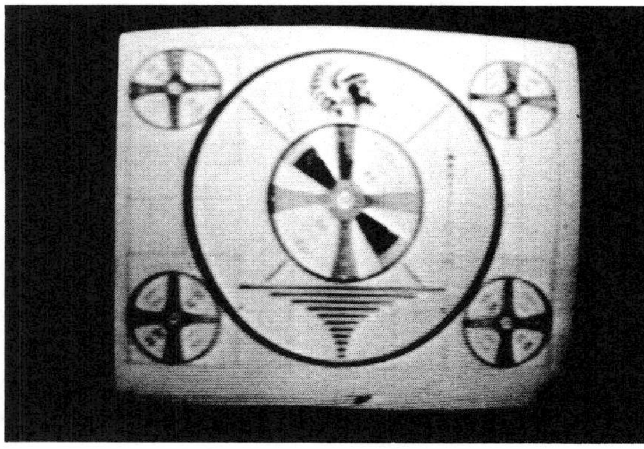

Fig. 11-20. Reduced width.

capacitor. In some of the more economical sets, a certain amount of nonlinearity should be expected and tolerated. Some sets have a separate *hori-*

Fig. 11-19. Mild horizontal-sync trouble (pulling).

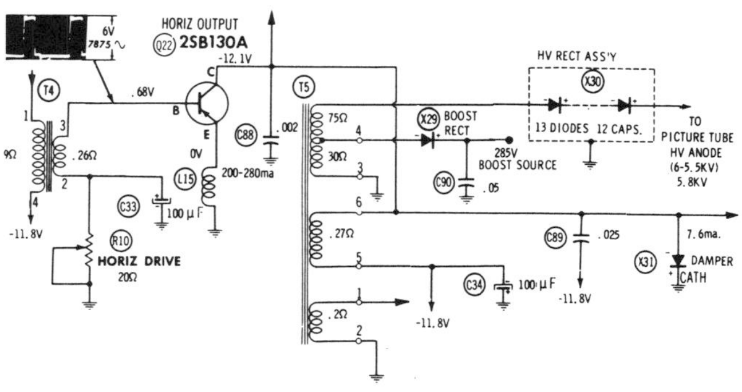

Fig. 11-21. A horizontal-output circuit.

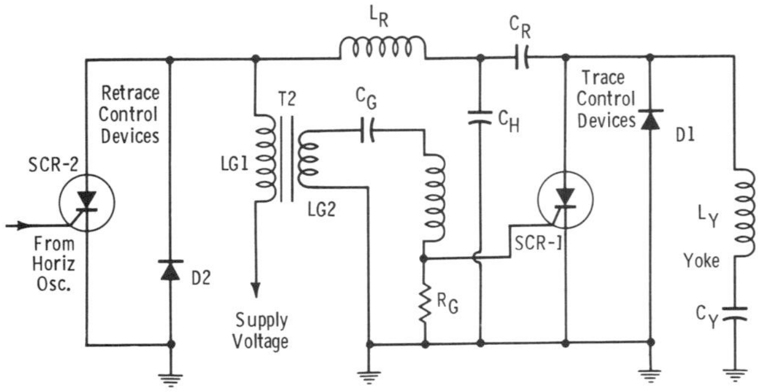

Fig. 11-22. An SCR horizontal-output circuit.

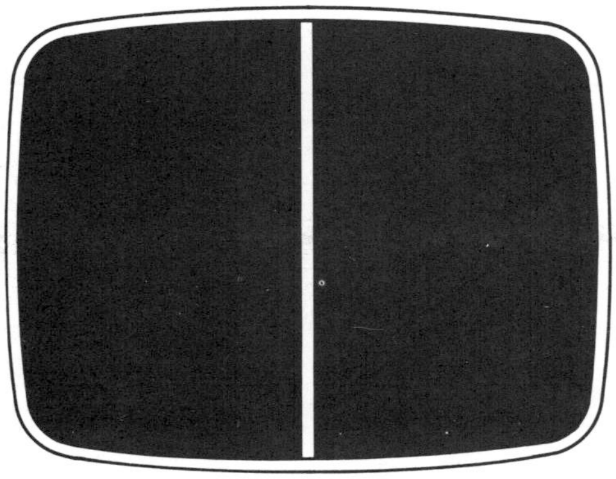

Fig. 11-23. No horizontal sweep.

zontal-linearity, or *efficiency,* coil for adjusting the picture.

In many cases, nonlinearity accompanies other symptoms such as loss of picture width. When you find such a case, you should concentrate on the other symptom and leave the linearity problem for the last. Often, eliminating the more serious problem will also correct the linearity.

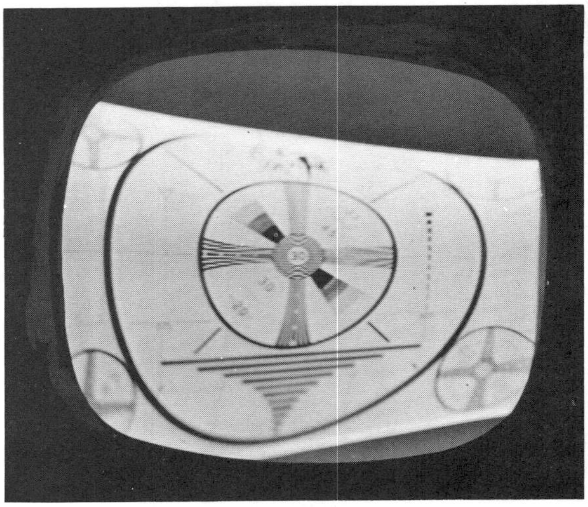

Fig. 11-24. Keystoning in the yoke (vertical coils).

Fig. 11-25. Horizontal nonlinearity.

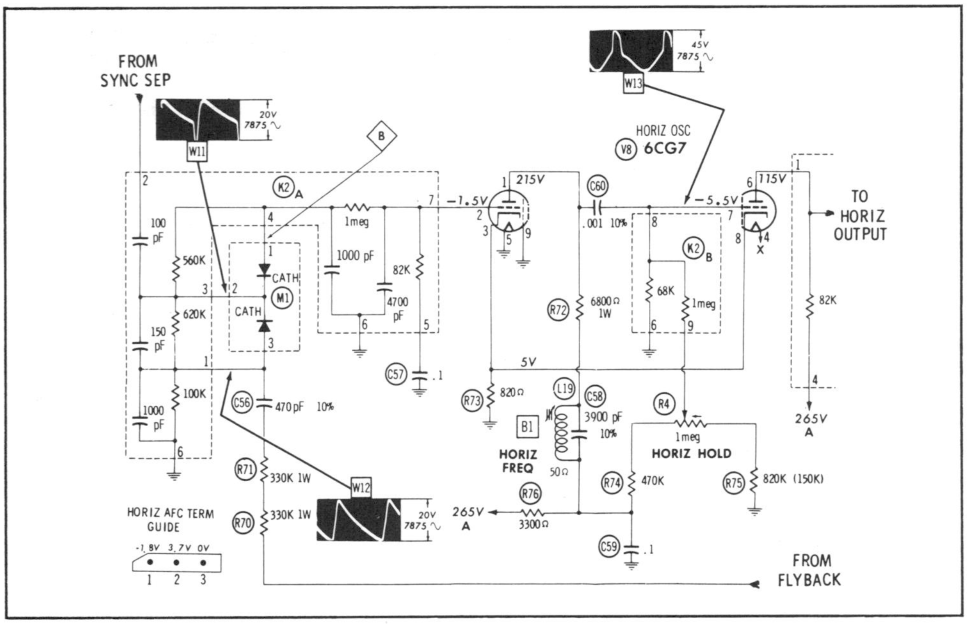

Fig. 11-26. A vacuum-tube afc-oscillator circuit.

11-11 Loss of the Scanning Raster

Since the high-voltage power supply is driven directly by the horizontal-sweep signal, loss of the sweep signal will knock out the high voltage. Without high voltage, there is no scanning raster. It is important to know that any horizontal stage from the afc or oscillator on can cause a high-voltage failure. Chapter 12 will be devoted to this problem.

11-12 Vacuum-Tube Horizontal-Sweep Circuits

The circuits used in the horizontal-sweep sections of tube-type receivers are like the transistor ones in many ways. Fig. 11-26 shows a typical vacuum-tube afc-oscillator section. Notice that the afc circuit is almost the same as the one shown in Fig. 11-18. While the tube circuit does not have a horizontal driver stage at all, it does have a two-tube oscillator circuit. The only real difference is that if the driver in a solid-state set fails, the oscillator might keep operating. If either tube in Fig. 11-26 fails, the oscillator quits. The symptoms in either case would be the same.

Fig. 11-27 shows a vacuum-tube horizontal-output circuit. Compare this circuit with the transistor circuit shown in Fig. 11-21. Notice that except for the voltages given, they are quite alike.

11-13 Troubleshooting Horizontal-Sweep Circuitry

In the horizontal afc-oscillator and driver stages, you should troubleshoot just as you would in any other circuit. That is, you should measure the operating voltages and compare them to the values given on the diagram of the set. If that does not locate the cause of the trouble, use an oscilloscope to view the circuit waveforms, and compare them to the ones given on the schematic. One of these two techniques will give you enough information to allow you to find the defective portion of the circuit. Testing the components in the defective portion of the set will reveal the bad one. In case out-of-circuit tests do not reveal the defective component, substitute known-good ones.

The horizontal-output stage is a somewhat different story. It is OK to check the waveform at the *input* of the horizontal-output transistor or tube. But you *must not* make measurements or view waveforms in the output part of the circuit unless you have special test instruments for that purpose.

It is not only dangerous for you due to the high energy levels, but you may also damage both the set and your instruments. However, since there are really very few parts in this stage, you can always turn the set off and check each one individually.

A good general rule is to be very careful in deciding whether the problem is in the output stage or in one of the earlier ones. Viewing the waveform at the input of the output stage will be very helpful in making this decision. If it is normal, the problem is in the output stage or beyond. If it is not normal, the problem is before the output stage. To illustrate the troubleshooting procedures consider a few examples.

Suppose that you have a set that has no vertical sweep. The symptoms are shown in Fig. 11-8. The sound is normal, and the vertical-sweep circuit is

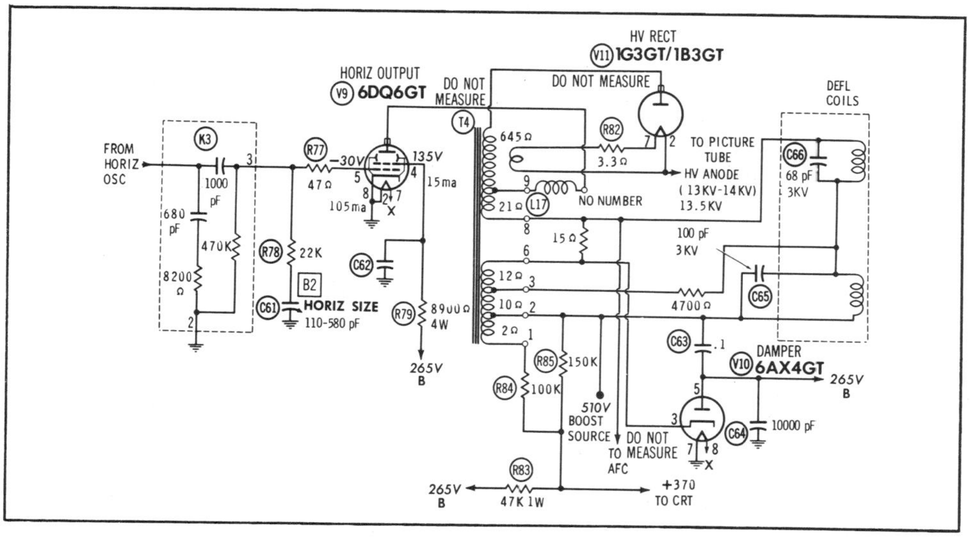

Fig. 11-27. A vacuum-tube horizontal-output section.

like Fig. 11-7. Check the operating voltages with a meter. Suppose you find that the collector-to-emitter voltage of Q16 is almost 12 volts (it should be about 9 volts according to the diagram). Check the emitter-to-ground voltage—it is zero. Looks like an open transistor; better check it. A quick test proves transistor Q16 to be open collector to emitter. Replacing it restores the sweep, but the picture is out of sync; probably just needs to have the vertical-hold control adjusted. If the hold control does not seem to have any effect, check the oscillator circuit. The base voltage measures zero. The oscillator is running, so the transistor is probably not too bad. Now think—if C19 were shorted, the base voltage would be zero. Yes, that is it, C19 is shorted. Replacing it with a new one corrects the sync problem. Now the picture is nonlinear—maybe it only needs adjustment of the linearity control. Yes, that's all it was; the picture looks fine now. Two such problems in the same circuit *is* unusual.

Here is another set. This time the picture is not wide enough (Fig. 11-20), and it is a little too tall. The horizontal circuits are like those given in Figs. 11-17 and 11-21. Use the scope and check the waveform at the base of the output stage (Fig. 11-21). It is not nearly as big as the one on the diagram. That probably means that the trouble is in the afc-oscillator section (Fig. 1-17).

Check the waveform at the driver base—it looks OK, so the cause must be a weak driver transistor. A new driver transistor fixed the width problem, but the picture is still a little too tall. Probably just needs its height control adjusted. There—that's better, it looks good now.

11-14 Review Questions

1. Explain in your own words what the function of the sweep-oscillator circuit is in both the vertical- and horizontal-sweep sections.
2. How are the two sweep oscillators different? How are they alike?
3. What are the symptoms if the vertical-sweep oscillator is not operating at the correct frequency?
4. How can you tell the difference between a sync problem in the sweep section and one in the sync section?
5. What are the basic types of failures that you can have in a horizontal-sweep section?
6. The circuit shown in Fig. 11-7 is out of sync. Which stage would you suspect of improper operation?
7. What could cause the symptoms shown in Fig. 11-6?
8. Describe vertical nonlinearity in your own words.
9. How do you recognize nonlinearity in a vertical-sweep waveform?
10. What are the basic types of failures that you can have in a vertical-sweep section?
11. The circuit shown in Fig. 11-17 is out of sync. Which part of the circuit would you suspect?
12. Refer to Fig. 11-5 for the following problem. All of the operating voltages are normal, and so is the waveform at the base of Q18, but the picture is severely nonlinear. The vertical-linearity control helps some but not enough. Which components would you check?
13. Refer to the circuit given in Fig. 11-7 for the following trouble. The symptoms are the same as in Question 12. The waveform at the base of Q15 is normal, but the one at the base of Q16 is nonlinear. Which components would you check?
14. In Fig. 11-7, the oscillator waveforms are all normal, but the waveform at the base of Q15 is very weak. What would be the picture symptoms? Which components would you check?
15. Describe horizontal nonlinearity in your own words, and tell what causes it.
16. Refer to the circuit given in Fig. 11-17 for the following problem. The voltage from collector to emitter of Q21 is zero. What would you suspect?
17. What would be the picture symptoms in Question 16?
18. The oscillator in Fig. 11-17 is operating at the wrong frequency. Otherwise, the circuit seems normal except that the horizontal-hold control has no effect at all. Which components would you suspect?
19. The waveform at the input of Fig. 11-21 is normal, but the set has reduced width. What would you do?
20. The waveform at the input of Fig. 11-21 is missing. What part of the set would you check next?

CHAPTER 12

Loss of the Raster

The scanning raster is composed of the white lines "painted" on the screen by the sweep circuits. The video information is superimposed on the raster. Any problem that prevents the raster from being the correct size and shape can be considered a raster problem. With this in mind, almost any raster problem can be classified as a sweep trouble. Many of these types of raster problems were discussed in Chapter 11. In this Chapter, only those *sweep* troubles which result in a loss of the raster will be considered. In addition, the high-voltage power supply and its effect on the

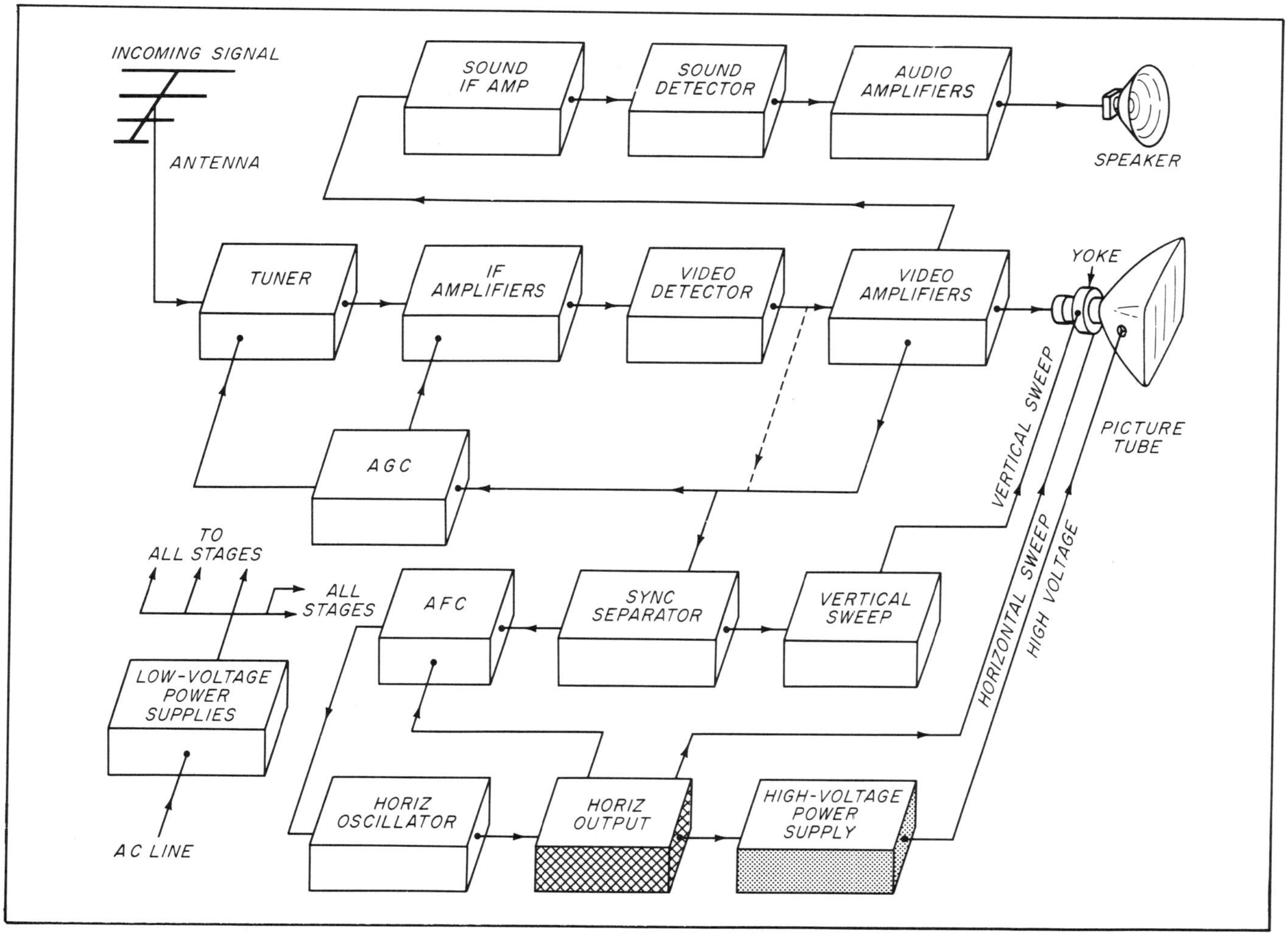

Fig. 12-1. Raster stages indicated in diagram of typical black-and-white receiver.

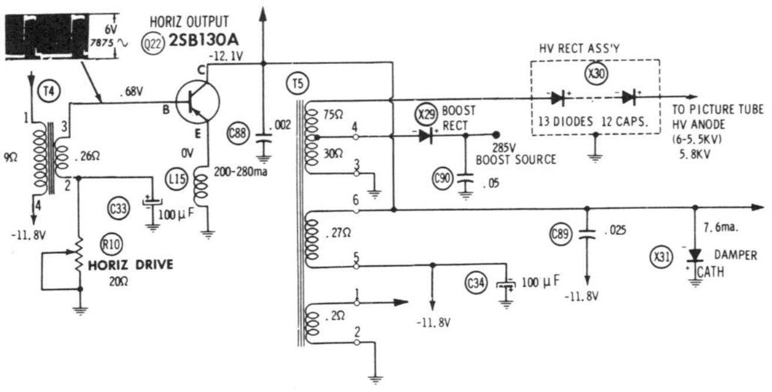

Fig. 12-2. A high-voltage power supply.

raster must be taken into consideration. Fig. 12-1 shows the blocks that are primarily concerned with the raster.

12-1 High-Voltage Rectifier Circuits

As it was noted in Chapter 11, the high-voltage power supply is driven by the horizontal-output stage. Fig. 12-2 shows a typical circuit. If there is no output from the horizontal-output stage, there will be no high voltage. The output from the horizontal-output stage can be absent if:

1. The horizontal oscillator fails.
2. The horizontal driver fails.
3. The horizontal-output stage fails.

Also, since the damper stage is used to produce the boosted B+ to operate the horizontal output, a bad damper can also kill the horizontal section.

The output of the horizontal-output stage is a series of high-energy pulses at the horizontal-sweep frequency. These pulses are stepped up through the horizontal-output (*flyback*) transformer. The high-voltage pulses at the output of the transformer are rectified by the high-voltage rectifier diode, and the resulting dc voltage is applied to the picture-tube anode. The high-voltage rectifier diode itself may be solid state, as shown in Fig. 12-2, or it may be a vacuum tube. A typical vacuum-tube high-voltage stage for a hybrid set is shown in Fig. 12-3. In a set that uses vacuum tubes only, the high-voltage rectifier will, of course, also be a tube. Fig. 12-4 shows a high-voltage section from a vacuum-tube set.

The value of the dc high voltage for a tv set ranges from around 6000 volts for a small black-and-white set up to 26,000 volts for a large-screen color set.

12-2 High-Voltage Trouble Symptoms

If the high voltage fails, then the picture-tube screen cannot light up. As a result, the symptoms that are associated with a high-voltage failure are normal sound and no raster. Fig. 12-5 shows the no-raster symptom.

Some failures in the high-voltage section can cause the high voltage to be a little weak but still large enouh to operate the picture tube. When this happens, the electron beam inside the picture tube is not attracted as strongly toward the screen. As a result, the beam moves more slowly. This slow-moving beam spends more time inside the yoke region of the tube. Therefore, the yoke has a greater effect, and the beam is *overdeflected*. This causes the picture to be too big for the screen. Fig. 12-6 shows such a case.

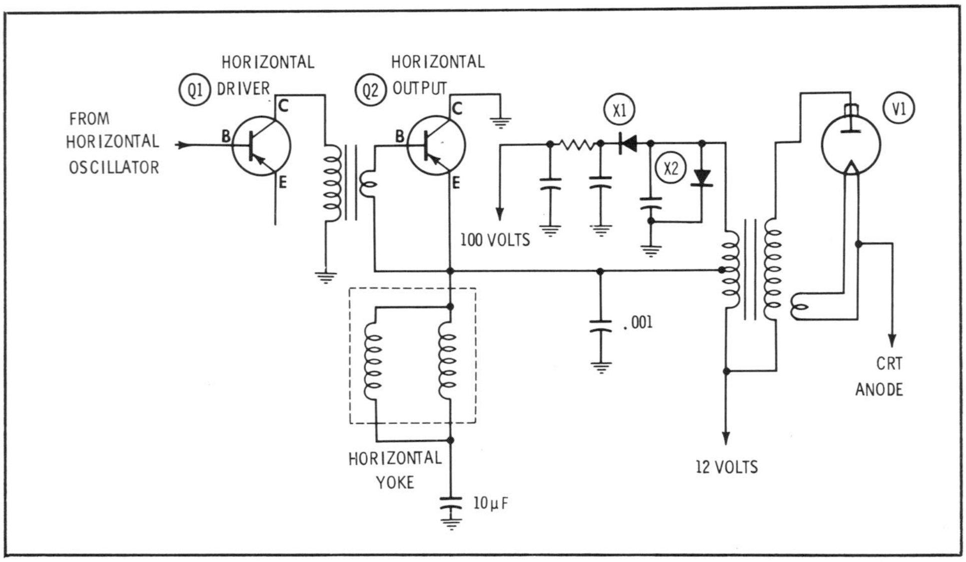

Fig. 12-3. A vacuum-tube high-voltage rectifier circuit.

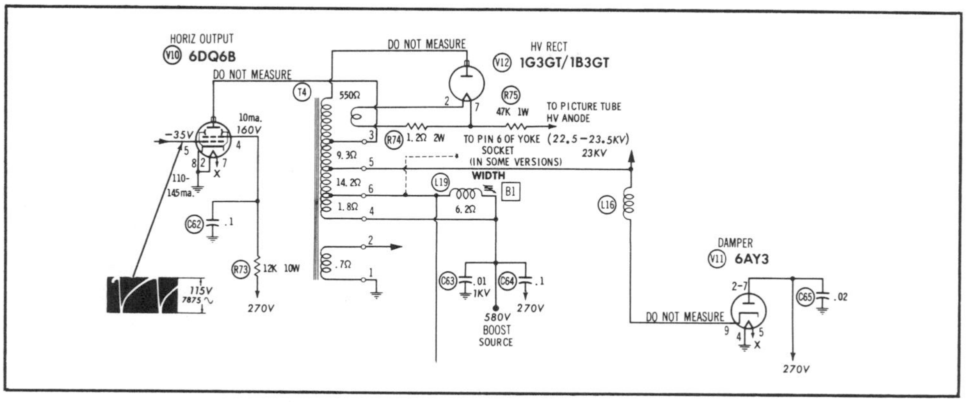

Fig. 12-4. High-voltage section in a vacuum-tube set.

In describing this symptom, it is generally said that the picture is *blooming*. Blooming is usually accompanied by poor focus. Perhaps the most common causes of weak high voltage and blooming are bad high-voltage rectifier diodes and failures in the high-voltage transformer.

12-3 High-Voltage Regulators

The high voltage is much more critical for a color picture tube than it is for a black-and-white set. Even small changes in high voltage can cause discoloration of a color picture. So, in a color set you will find a *high-voltage regulator* of some kind. There are three popular methods of regulating the high voltage in a color set. They are: the *shunt-regulator* method, the *horizontal-regulator* method, and the *horizontal-bias regulator* method.

A shunt regulator is used in most vacuum-tube sets and many hybrid sets. Fig. 12-7 shows the block diagram of this type of regulator. The regulator circuit monitors the value of the high voltage directly and holds it at a preset value. If the regulator fails, it can cause a total or partial loss of high voltage with the resulting no-raster,

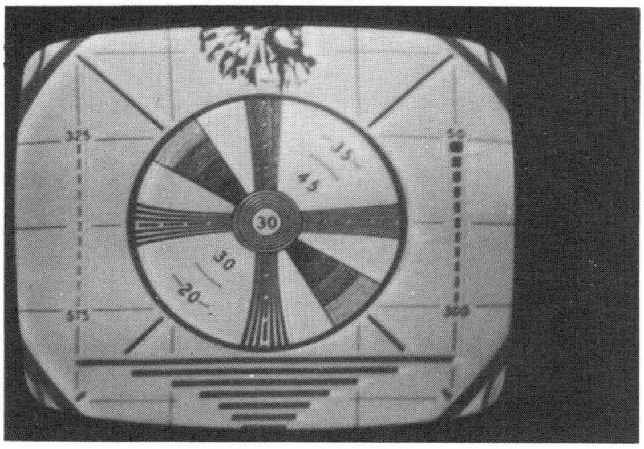

Fig. 12-6. Blooming caused by reduced high voltage.

Fig. 12-5. No-raster symptom.

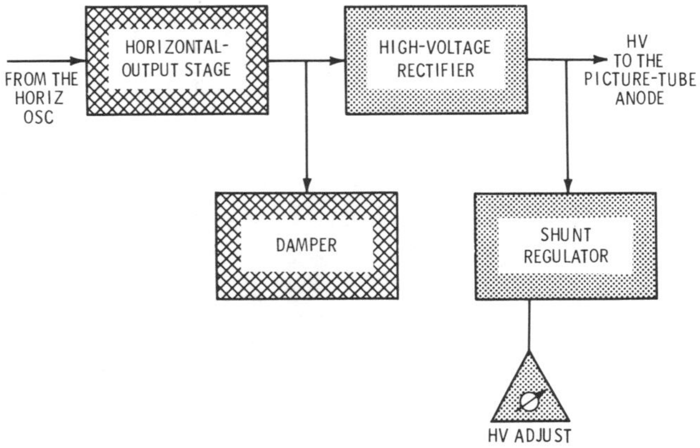

Fig. 12-7. Shunt-regulated high-voltage system.

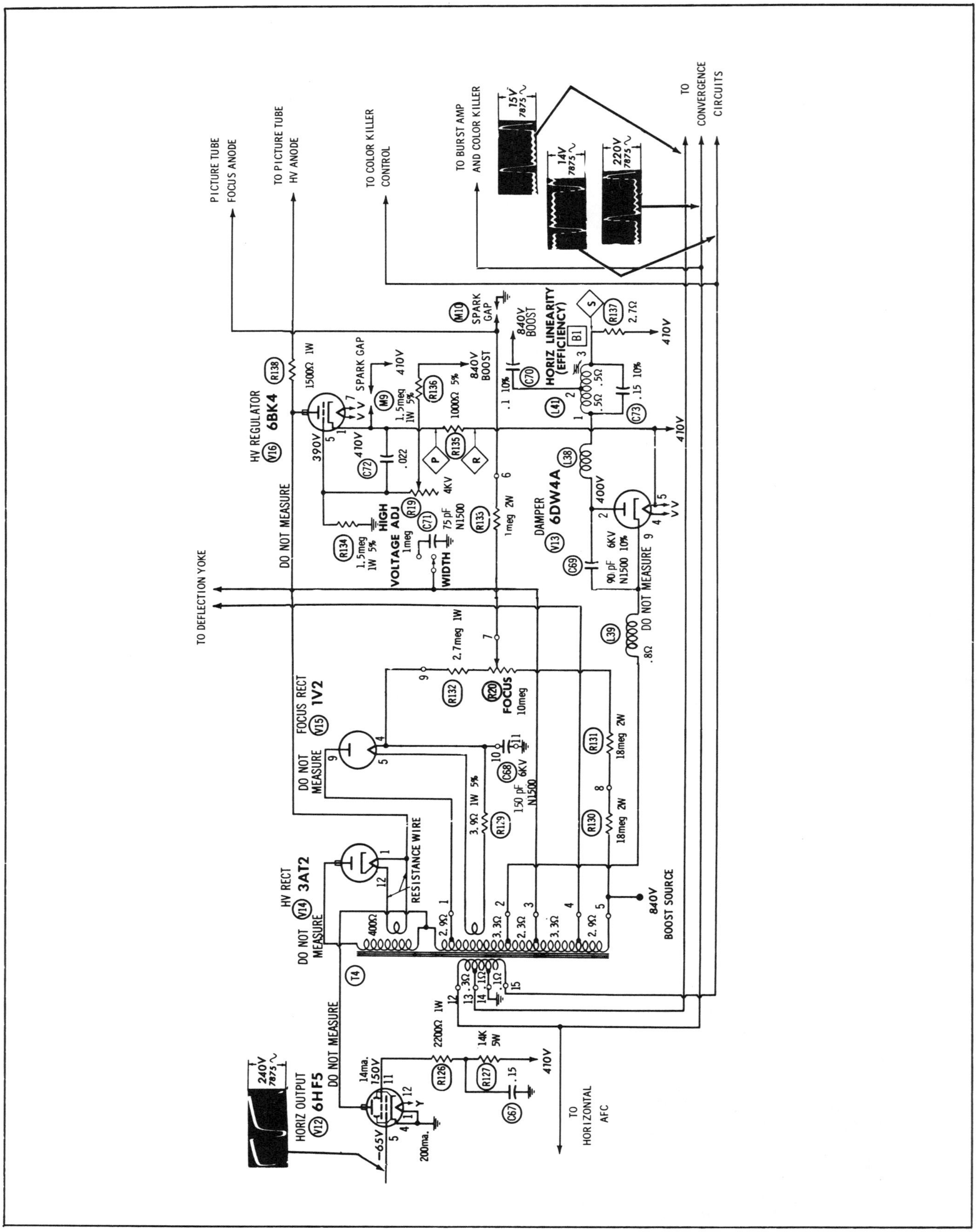

Fig. 12-8. A shunt-regulated high-voltage system.

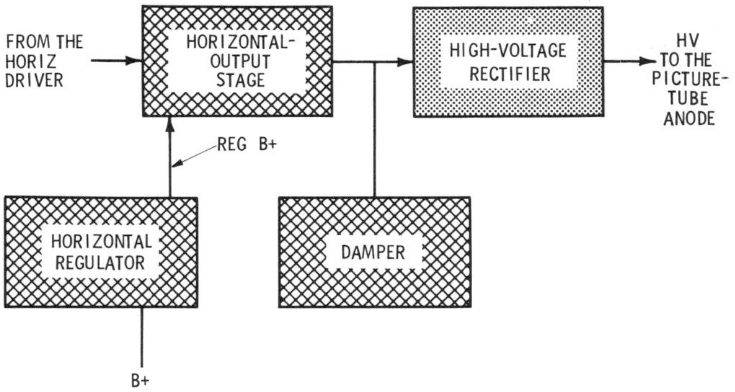

Fig. 12-9. A horizontal-regulator type of high-voltage system.

blooming, and discoloration symptoms described previously. On the other hand, it can also fail in such a way as to cause the high voltage to be too high. When this happens, the usual symptoms are a discolored picture, arcing, corona, and in extreme cases burn-out of other components. Fig. 12-8 shows the circuit diagram of a typical high-voltage shunt-regulator circuit.

It is worth mentioning at this point that X-rays may be generated when high-velocity electrons collide with certain materials. The amount of X-ray energy produced depends on the high-voltage value and the material bombarded. Television-set manufacturers design the sets to operate within established safety standards. The high-voltage value in a set should not be increased beyond the specified level. By increasing the high voltage, you may cause the production of levels of X-radiation that are beyond the established safety standards.

The second type of high-voltage regulator that is commonly used is the *horizontal-regulator* type. Fig. 12-9 shows a block diagram of this type of regulator. The horizontal regulator is based on the fact that the high-voltage output depends directly on the output of the horizontal section. By regulating the B+ voltage to the horizontal-output stage, the high voltage can be regulated. This type of regulator is used in many solid-state color sets. However, there is no reason that it cannot be used in hybrid or even vacuum-tube circuits. Fig. 12-10 shows the horizontal-regulator circuit from one solid-state receiver.

In the third type of circuit, regulation is achieved by controlling the bias applied to the horizontal-output stage. Fig. 12-11 shows an example of the circuit. In addition to the bias developed by the drive from the horizontal oscillator, there is also a bias developed by pulses that come from a winding on the flyback transformer through the pulse-coupling capacitor (0.001 μF).

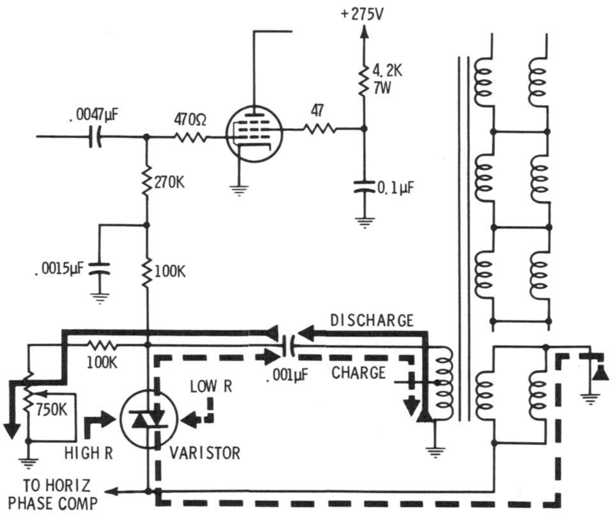

Fig. 12-11. Regulation by horizontal-output bias.

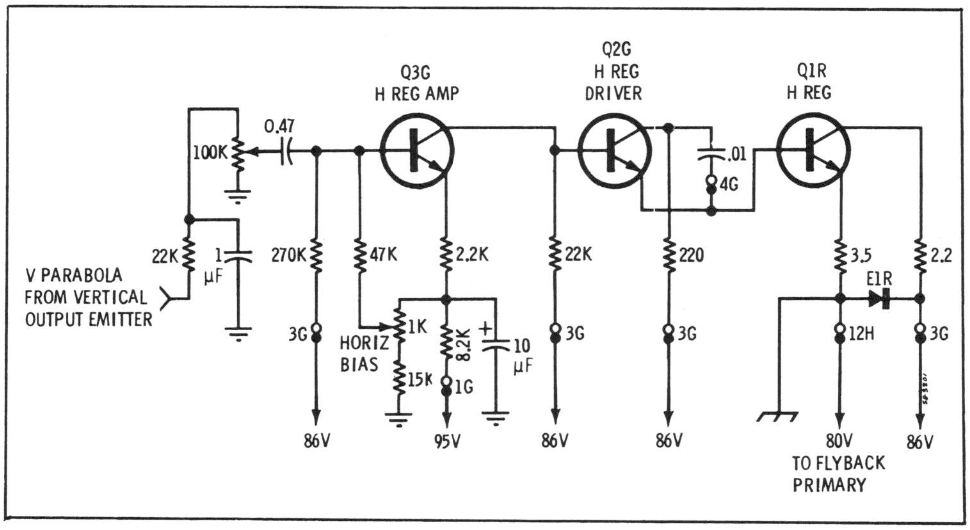

Fig. 12-10. Horizontal regulator for a solid-state color tv receiver.

These pulses appear across the horizontal-bias control and the 100K resistor in series with it.

There is another winding on the flyback transformer, which is used for sampling for the phase comparator. This winding is placed in series with a varistor, which is in series with the pulse-coupling capacitor. The pulse from the bias winding will be a positive-going pulse, while the pulse from the phase comparator winding will be negative-going. This causes a large pulse-voltage difference across the varistor. With a large voltage across the varistor, its resistance will be low. The coupling capacitor will charge during the retrace pulse time, as shown by the dashed line in Fig. 12-11. Due to the low resistance of the varistor with pulses present, the charging time constant is short. The voltage developed across the capacitor will act as a bias source for the grid of the horizontal-output tube. During line time, the capacitor discharges (as shown by the unbroken heavy line in Fig. 12-11) through the 100K resistor and bias control. At this time, the varistor has high resistance. The discharge time constant is quite long and thus maintains the developed fixed bias for the horizontal-output stage.

With the high voltage present, possible danger of X-ray radiation and fire within the high-voltage section exists. Each manufacturer gives very special attention to the design of the high-voltage cage and the placement of parts and leads within the cage. Sometimes tubes with leaded-glass envelopes are used in the horizontal section. Under no circumstances should the service technician take it upon himself to change the safety features designed into the cage. The high-voltage shielding must be restored to its original condition with doors fastened before you leave the set.

12-4 Focus-Rectifier Circuits

Some color picture tubes require as much as several thousand volts on the focus electrode to get a sharp spot. For this reason, some sets have a special circuit to produce the focus voltage. Fig. 12-12 shows one of the focus-rectifier circuits in use today. If the focus rectifier fails, the set will be badly out of focus, and in many cases the high voltage will be affected as well. In other cases, a total failure of the focus can leave the high voltage normal but still produce a no-raster symptom.

The use of a focus rectifier like the one shown in Fig. 12-12 is not the only way focus voltage can be produced. Sometimes a tapped voltage divider is connected directly across the high-voltage output. Or, a solid-state voltage multiplier may be used together with a winding on the horizontal-output transformer.

12-5 Measurements in the High-Voltage Section

Because of the high voltage and energy levels that are present, measurements in the high-voltage section require special attention.

The high-voltage values should be measured only with an accurate high-voltage probe. These probes are available for the better multimeters. They are a must for a fully qualified technician. Some of the more modern probes have a built-in meter and are designed for only high-voltage measurement. These handy probes make a nice addition to the service caddy and provide an accurate means for the service technician to measure and adjust the high voltage to the manufacturer's specifications in the home. In some systems, the high voltage may vary several thousand volts as the brightness level and contrast are increased. For this reason, most manufacturers stress that this adjustment should only be made with the brightness and contrast set at minimum (zero beam current).

You should not attempt to view any of the high-voltage points with an oscilloscope unless the scope has a high-voltage test probe. The high-energy pulses present in the high-voltage section can easily damage the oscilloscope. However, waveforms can be viewed by clipping the scope lead onto the insulation of the wire. There are usually some waveforms that can be viewed, and these are normally shown on the circuit diagram.

Test points are frequently provided by the manufacturers for monitoring the current through the

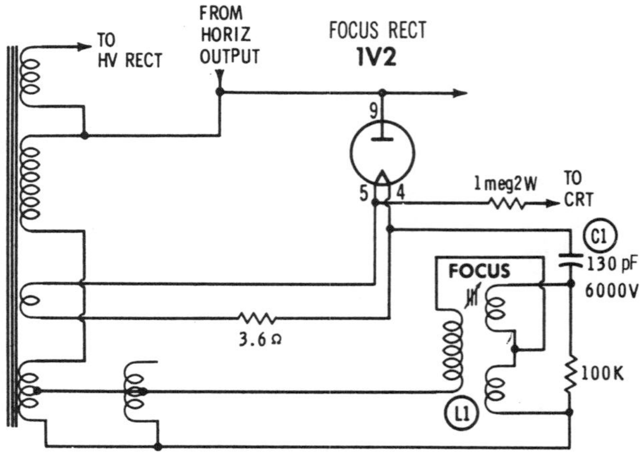

Fig. 12-12. Typical focus-rectifier circuit.

horizontal section. Any trouble in the horizontal section will cause variations in the current or high-voltage limits specified by the manufacturer. The manufacturer normally gives specific details about how the measurements are made and how the test equipment must be connected.

Probably the most useful troubleshooting technique for use in the high-voltage section is to turn off the set and check the suspected parts one at a time with an ohmmeter. This technique will not, of course, identify shorted turns in one of the coils. Usually the high energy levels present *burn* defective components rather badly. They are, therefore, easy to spot.

12-6 Video Blanking Problems

In addition to high-voltage or focus problems that cause loss of the raster, video blanking can produce very similar symptoms. Any video or picture-tube problem that causes the picture tube to stay completely cut off (blanked) will result in a no-raster condition.

12-7 Review Questions

1. What are the symptoms usually associated with a complete failure in the high-voltage section?
2. Which one of the following is a high-voltage problem?

   (A) No sound (C) No sync
   (B) No raster (D) No picture
3. The horizontal output in Fig. 12-3 seems to be OK. There is no high voltage at the picture-tube anode, but replacing V1 does not seem to help. What would you do next?
4. There is no raster on a set that uses the circuit of Fig. 12-2. A quick check shows the waveform at the base of Q22 to be normal. List the things you would check next.
5. Describe in your own words what a blooming picture looks like. What is the basic cause of a blooming symptom?
6. What are the three types of high-voltage regulators, and how do they work?
7. What are the symptoms of a failure in a high-voltage regulator?
8. If the damper shown in Fig. 12-4 goes out, will the high voltage be affected? Why?
9. What does the focus rectifier do in a color tv set?
10. How is the high-voltage section in a color set different from that in a black-and-white set?
11. What would be the symptoms if X30 in Fig. 12-2 were to fail?

CHAPTER **13**

Low-Voltage Power-Supply Troubles

13-1 Symptoms of Low-Voltage Power-Supply Failure

The low-voltage power supplies in a television set furnish the operating voltages for all of the various stages. Many sets have only one actual low-voltage power supply. When this single supply fails, the whole set goes off. In this kind of set, a low-voltage failure would be indicated if there was no raster and no sound.

Some sets have two or more independent low-voltage power supplies. If a single supply fails in this kind of set, then almost any symptoms can occur, depending on which stages that supply was operating.

All line-powered sets have one or more low-voltage power supplies operated from the 60-Hz ac line. In addition to these 60-Hz power supplies, many sets have other low-voltage power supplies that are operated from the 15,750-Hz horizontal-output circuit. We will consider the sweep-powered circuits later, in Section 13-5.

13-2 The AC Distribution System

The ac distribution system is the wiring that gets the 120-volt, 60-Hz house current from the wall plug to the low-voltage power supply. While the exact circuitry varies from set to set, most receivers have some things in common. All sets have a wall-outlet plug and a line cord. They all have an interlock that disconnects the line when the back cover of the set is removed. They all have an on-off switch; and, almost all sets have some type of ac protective device. This protective device may be a fuse or a circuit breaker. Fig. 13-1 shows a typical television ac distribution circuit. The resistor from one side of the line to ground (chassis) is put in to help eleminate danger from shock. Some sets use capacitors to provide the same protection.

A feature comon to modern color sets is the automatic degaussing circuit. This circuit is shown in Fig. 13-2. When the set is first turned on, the degaussing coil demagnetizes the picture tube, allowing better color quality. After a short time, the thermistors change value so that the current through the degaussing coil decreases nearly to zero. This cycle is repeated each time the set is turned on.

Many of the more modern black-and-white and color tv sets are being provided with a special type of line cord that allows the set to be plugged into the ac house outlet in only one direction. The plug on this type of cord has one prong wider than the other. Most homes built within the last few years use ac receptacles designed for this type of plug. The wider slot of the ac receptacle is connected to ground through the house wiring. Some sets designed for commercial installation use a three-wire ac line cord. Both types of cord ensure proper grounding of the chassis and any exposed metal parts. Any set owner not having a modern type of ac receptacle for his set should be en-

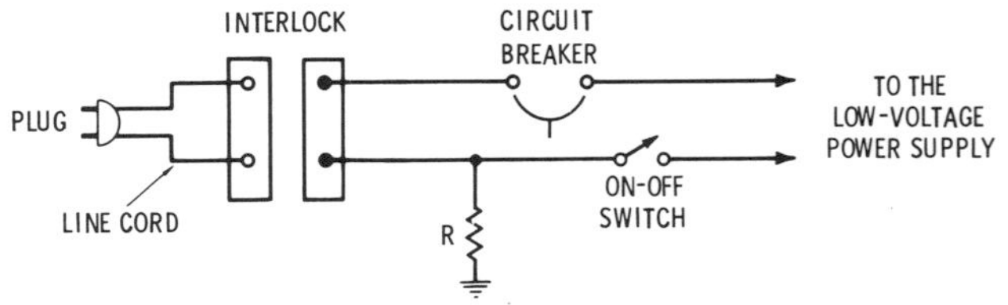

Fig. 13-1. Typical ac distribution circuit.

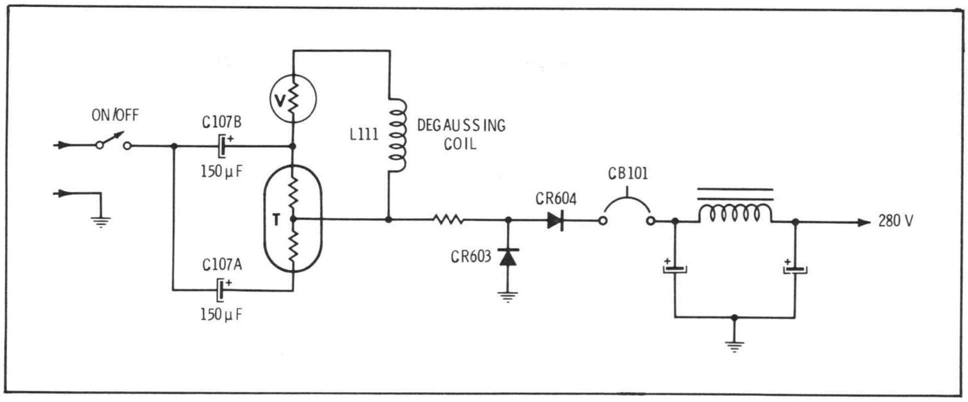

Fig. 13-2. Typical automatic degaussing circuit.

couraged to have an electrician install the modern receptacle. Under no conditions should the technician alter or replace the ac line cord in such a way as to destroy this safety feature.

While the ac distribution section is considered part of the "low-voltage" system, keep in mind that this section has perhaps the highest shock hazard of any part of the set. *Always be very careful when working in the low-voltage sections.*

It is always a good idea to use an *isolation transformer* when servicing a set. Many sets (even some with power transformers) have one side of the power line connected to the chassis ground. Failure to use an isolation transformer can result in a dangerous shock hazard.

13-3 Low-Voltage Power-Supply Circuits

The low-voltage power supplies in a television receiver are designed to satisfy the particular requirements of that set. As a result, there are almost as many different power-supply circuits as there are models of sets. But, as in the case of the ac distribution system, there are some basic elements that a power supply must have. Fig. 13-3 shows power-supply requirements in block-diagram form. Not every set will have all of these blocks, but every set will have some of them. Let us go through the blocks one at a time.

As you have already seen, every set that operates from wall current has an ac distribution circuit. The purpose of this circuit is to get the electricity from the wall outlet into the set.

Most television receivers have a power transformer. The purpose of the transformer is to convert the 120-volt line current into the voltages required in the set.

Some sets are designed to work directly from the line and do not have a transformer. These sets work as well from 120 volts dc as they do from 120 volts ac. For this reason, they are commonly called ac-dc sets. The fact that they will work on dc is of little value since almost all of our house current is now ac.

The rectifier circuit is an essential part of any power supply. Its job is to convert the ac voltage into a dc voltage. There are four popular circuits that are used most of the time in tv power supplies. Fig. 13-4 shows these circuits. The first one is called a *half-wave rectifier,* and its main advantage is that it is very economical to build. On the other hand, its output is hard to filter. It is used most often with circuits that do not require too much filtering.

The second circuit is called a *full-wave rectifier.* Its output is easier to filter, but the circuit requires a center-tapped transformer. It is probably used more in television than any other type of rectifier circuit.

The third circuit is called a *bridge rectifier.* Its output is easy to filter, and it does not need a center-tapped transformer, but it is more expensive than the other two circuits. Bridge rectifiers are used quite a lot in television sets, but not as much as the other type of full-wave rectifier circuit (Fig. 13-4B).

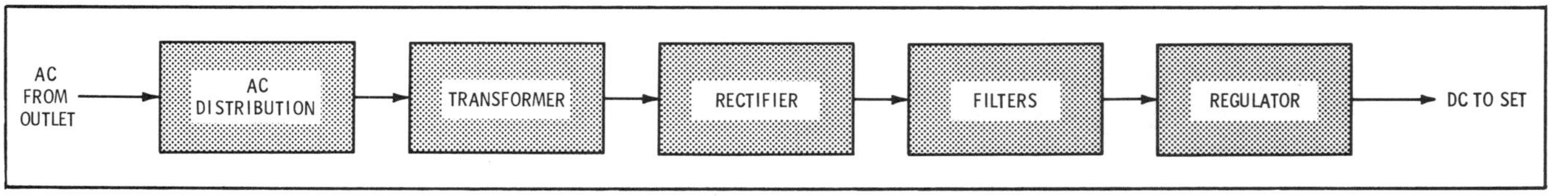

Fig. 13-3. Basic parts of a power supply.

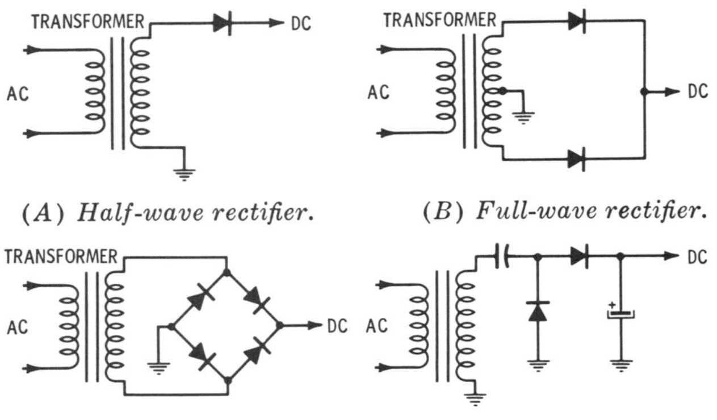

(*A*) *Half-wave rectifier.* (*B*) *Full-wave rectifier.*

(*C*) *Bridge rectifier.* (*D*) *Half-wave doubler.*

Fig. 13-4. Common rectifier circuits.

The last type of power supply is of the half-wave type, but it produces higher output voltages than the first one does. Under ideal conditions, this circuit can produce twice as much voltage as a simple half-wave power supply. Because of this capability, it is called a *half-wave voltage-doubler* circuit.

The output of any rectifier is a fluctuating dc current. Filter circuits are used to remove the fluctuations from the current. This leaves only smooth dc for use in the set. Fig. 13-5 shows the two most common types of power-supply filters. The RC filter is used most often because it is cheaper to build. The LC filter, while being more costly, does a better job of filtering out the ac component.

When very accurate control of the power-supply voltage is needed, a regulator circuit is used. There are two common types used in tv sets. They are the zener-diode circuit and the transistor regulator circuit. Fig. 13-6 shows a simple power supply with a zener-diode regulator. In this circuit, the zener diode (Z1) simply keeps the output voltage from rising above 12.6 volts. If the zener diode

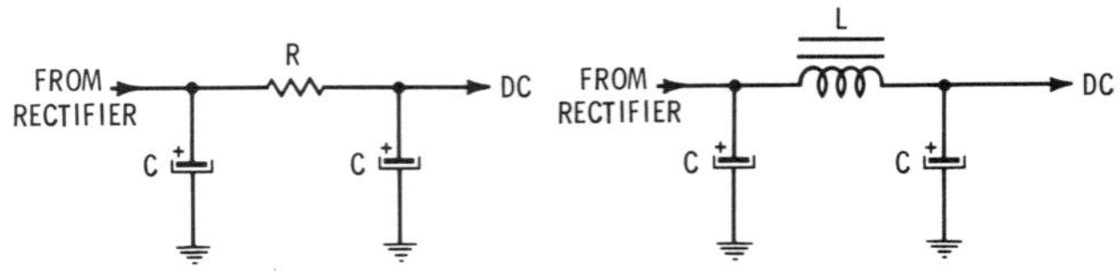

(*A*) *RC filter.* (*B*) *LC filter.*

Fig. 13-5. Filter circuits.

fails, it can cause the voltage to be either too high (zener open) or too low (zener shorted).

The power supply in Fig. 13-7 has a more complicated regulator of the transistor type. In this circuit, transistor Q2 samples the output through the 4.7K and 5K resistors. If there is a difference between the base voltage and the emitter voltage, the difference is amplified and fed to Q1, which in turn corrects the output voltage.

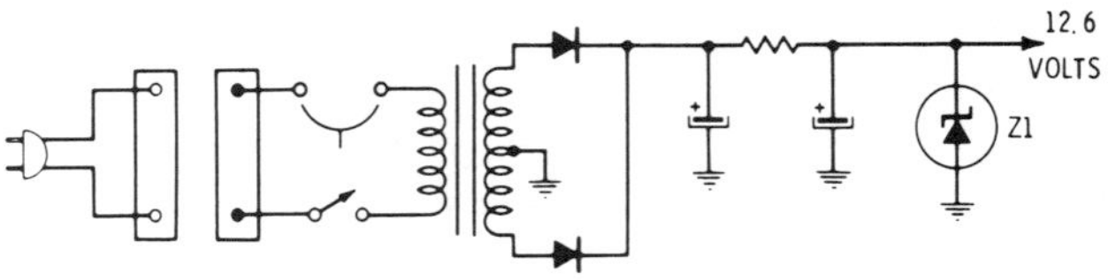

Fig. 13-6. A zener-regulated power supply.

Perhaps it is worthwhile at this point to examine one or two actual television power-supply circuits.

Fig. 13-8 shows the power supplies for a typical large-screen solid-state set. Notice that the ac distribution circuit on the left is very similar to the one previously considered.

The transformer has three secondary windings. The lower winding (as shown in the schematic) is used to light the picture-tube heater and the dial lamp. The center winding has two rectifier circuits connected to it. The lower rectifier is a full-wave circuit using diodes X504 and X505. The filter for this circuit is C506A and L502, and it is an LC

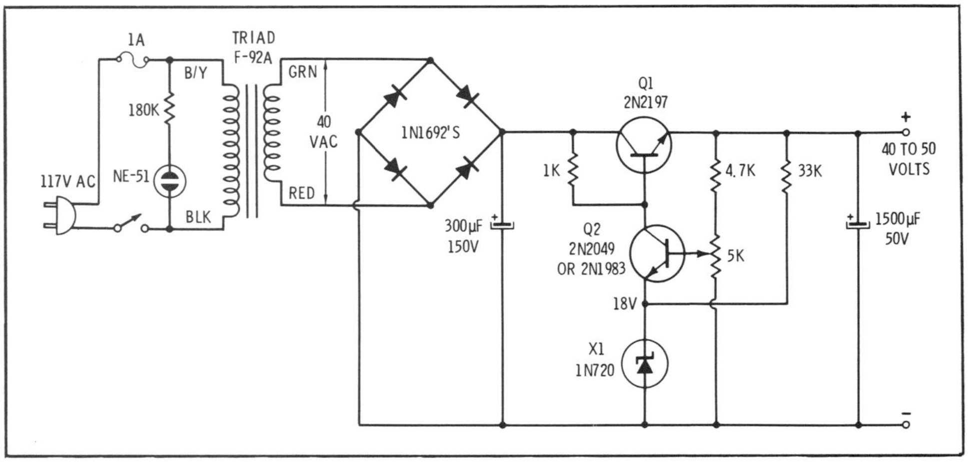

Fig. 13-7. Transistor-regulated power supply.

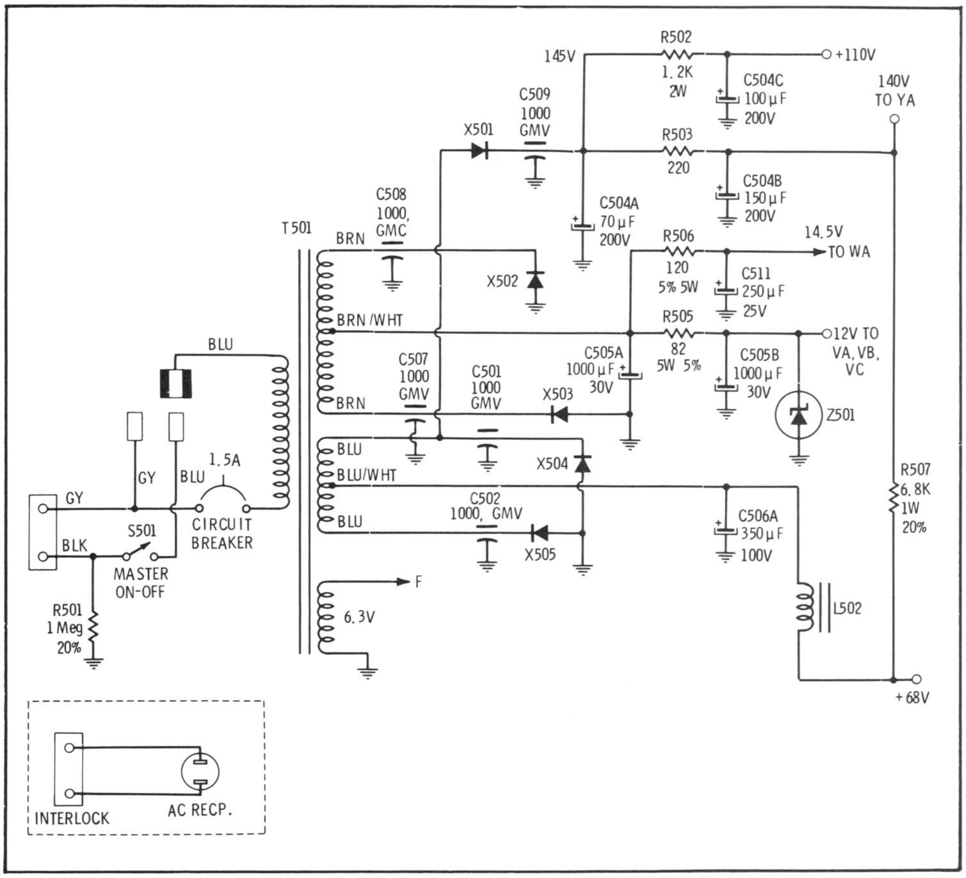

Fig. 13-8. Power supply for a transistor set.

type of filter. There is no regulator for this supply. The supply voltage is +68 volts.

The upper rectifier connected to the center transformer winding is a half-wave circuit (X501) with two filters. The outputs of this circuit are 110 volts and 140 volts, respectively.

The upper secondary winding supplies a full-wave rectifier (X502 and X503) and two filters. The upper filter branch delivers 14.5 volts and is unregulated. The lower filter branch provides 12 volts and has a zener-diode (Z501) regulator.

The power supply shown in Fig. 13-9 produces three different regulated outputs. The rectifier is the bridge type, and the regulator is a transistorized electronic one.

In many modern sets (usually the larger ones), a constant-voltage transformer is used to regulate the 60-Hz line voltage at about 100 volts. There are several types of constant-voltage (or voltage-regulating) transformers. The simplest type involves a transformer that saturates when the output voltage reaches a certain value. These transformers can effectively regulate line voltages from 100 volts to about 125 volts and maintain a relatively constant output.

13-4 Troubleshooting Low-Voltage Power-Supply Problems

There are three main types of problems that arise from low-voltage power supplies. They are:

1. Complete loss of voltage
2. Reduced low voltage
3. Hum

We will examine these types of problems one at a time starting with a complete loss of voltage.

If the set you are working with has only one power supply and it fails, then the whole set will be dead. That is, there will be no raster and no sound. Power-supply failure can be confirmed simply by measuring the output of the power supply with a voltmeter.

It is a good idea to measure the dc voltage across one of the filter capacitors with a dc voltmeter. If the dc voltage is zero, check the ac voltage at the input to the transformer. If the ac voltage is zero, then the problem is in the ac distribution circuit. If the ac is all right and the dc output is zero, then the problem is in the transformer, rectifier, or filter. Since there are usually only a few

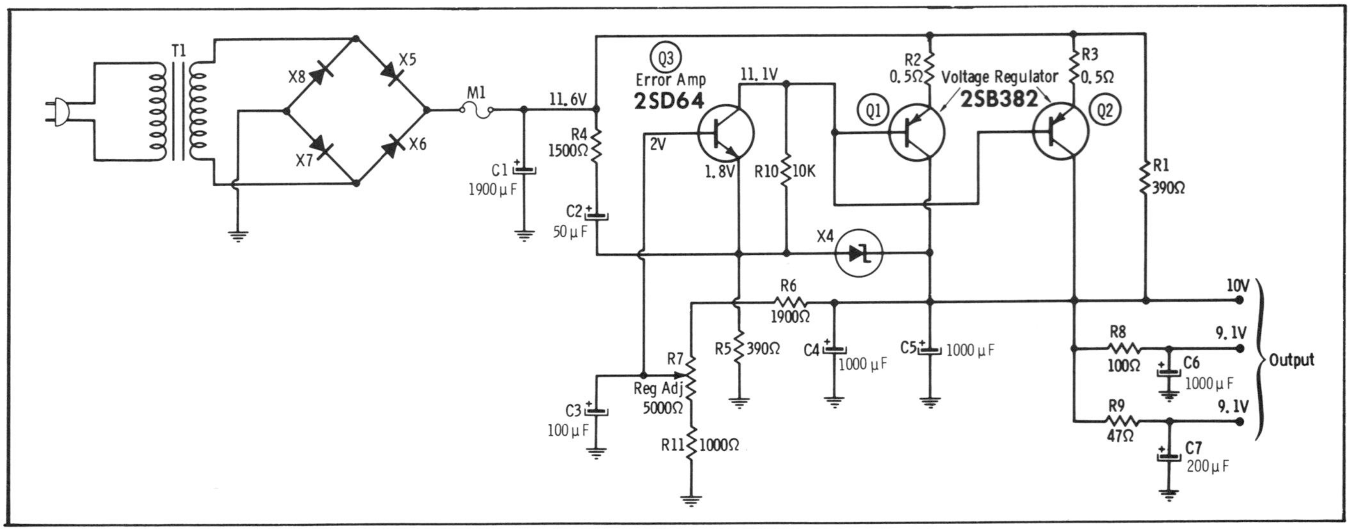

Fig. 13-9. Regulated bridge-type power supply.

parts in a power supply, each component can be checked individually if necessary.

A short circuit (or a near short circuit) is a fairly common problem that results in a tripped circuit breaker or a blown fuse. With the set unplugged, discharge the filters by temporarily shorting the power-supply output to common. Then check the resistance from the output to common with an ohmmeter. The resistance (after the filters charge) should be about 10,000 ohms or more. If it is not, check the circuit for shorts.

Many power supplies use copper-wire filament fuses (and other so-called *shaded* components). These parts *must always be replaced with exact* replacement components—no substitutions! When you are in doubt, always refer to reliable service data before making a substitution.

If the set has several power supplies and one of them fails, then only the part of the set depending on that supply will be dead. For instance, the power supply shown in Fig. 13-8 is actually three power supplies with five output voltages. These voltages are divided among the stages as follows:

140-volt output ..... video output stage only
110-volt output ..... audio output stage only
68-volt output ..... vertical- and horizontal-output stages only
14.5-volt output .... horizontal-driver stage only
12-volt output .... remainder of set

Suppose that R502 burned out. This would kill the 110-volt output, and the sound would be dead. As soon as you checked the audio output stage, you would find no collector voltage. Then trace the voltage back to the supply, where the trouble would be found.

If X501 failed, both the video and sound would be dead. You would find the trouble in the same way.

On the other hand, if the 14.5-volt output were out (say R506 burned out), then the horizontal driver would not work, and there would be no raster but the sound would be OK. Again, find the trouble by checking the horizontal-sweep section. You would find no voltage in the driver and trace the lack of voltage back to the power supply.

If the output of a power supply is low, the classic symptom is a shrunken picture like the one shown in Fig. 13-10. Sometimes, only the width is reduced. In extreme cases, both width and height can be affected. In troubleshooting this problem, you should check the voltages in the affected sections (horizontal and vertical), and then check

Fig. 13-10. Symptom of reduced low voltage.

Fig. 13-11. Hum in the video.

them in the power supply. Comparison with the values shown on the circuit diagram would reveal any reduced value. Reduced voltages are most frequently caused by either defective power-supply diodes or leaky filter capacitors.

Filter capacitors are also the most frequent cause of hum in the sound, video, or raster. When there is too much hum in the sound, check the filter capacitors in the power supply that operates the sound circuits.

Hum in the video appears as black horizontal bars in the picture. Fig. 13-11 shows this symptom. In this case, check the filter capacitors in the power supply that operates the video circuits.

Hum in the raster appears as bending; the sides of the raster have a sinusoidal curve to them. Hum can get into the raster through the stages that operate the horizontal-sweep circuits. Therefore, suspect the power-supply filter in the supply that furnishes voltage to these circuits.

An open filter capacitor can usually be tracked down easily by looking at the ripple voltage with an oscilloscope. When a capacitor is suspected of being open, a new one can be temporarily connected across it with the power off. Then the set is turned on to see if the problem has been cured.

13-5 Secondary Power Supplies

So far, we have assumed that the source of energy for our power supplies would be the ac distribution system. In many instances, this is the case, but it is not always.

When the energy source is some other circuit in the set, we consider the power supply to be a *secondary* one. A good example is the picture-tube high-voltage power supply, which receives its energy from the horizontal-sweep circuits.

Many set manufacturers use secondary supplies to provide operating voltages to other sections of the set. Fig. 13-12 shows a more or less typical secondary supply circuit. In this circuit, three half-wave rectifiers are operated from the horizontal-output stage. They provide voltages of 24, 33, and 235 volts to operate the vertical-sweep, sound, and perhaps other sections of the set.

One advantage of secondary supplies is that the ripple to be filtered out is at the horizontal-sweep

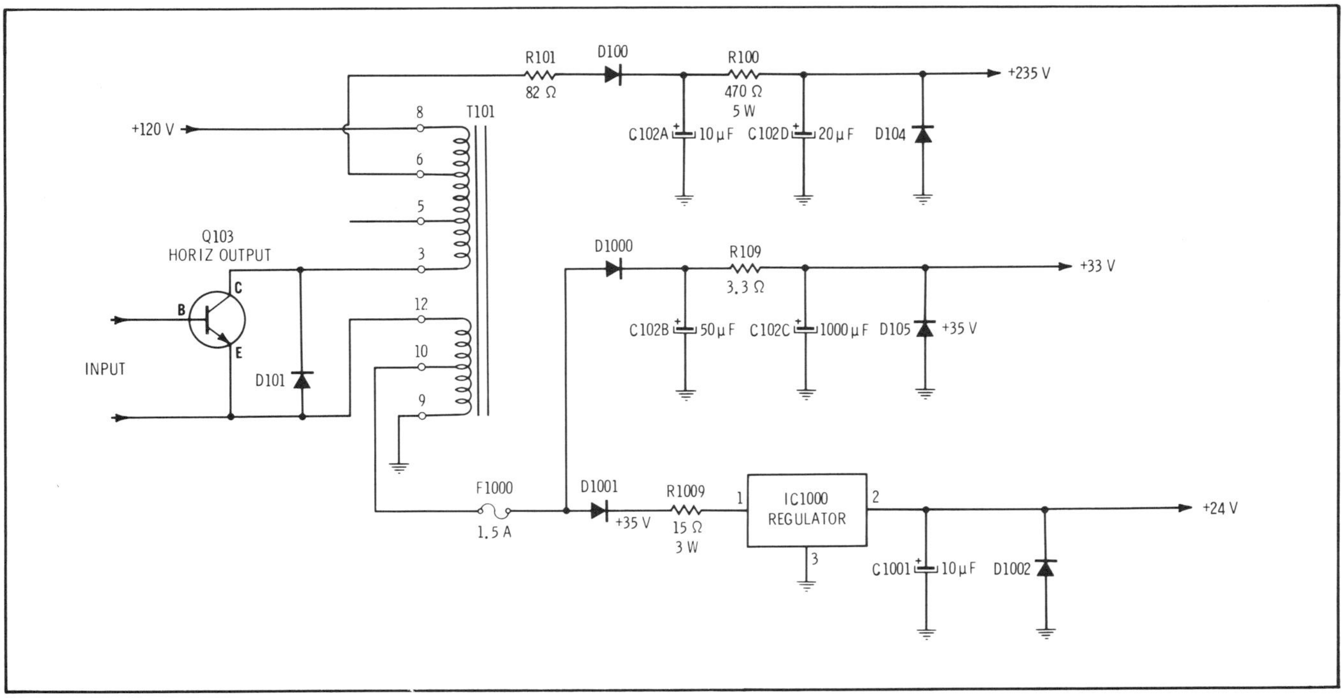

Fig. 13-12. A secondary power supply.

frequency of about 15,750 Hz. This makes filtering easier, and, in general, smaller filter capacitors can be used.

13-6 Review Questions

1. Examine the circuit diagrams at the back of this book, and answer the following:
   A. Which of the sets have only one power supply?
   B. Are any of the power supplies regulated? How?
   C. Do all of the sets have a transformer in the power supply?
   D. What type of filter circuits does each set have?
2. What would be the set symptoms if X504 in Fig. 13-8 were to fail?
3. A set that uses the power supply shown in Fig. 13-8 is dead. Even the dial lamp is out. What would you suspect first?
4. If the dial lamp in Question 3 is on, would you bother to check the circuit breaker? Why?
5. A set that uses the power supply shown in Fig. 13-8 has the symptoms shown in Fig. 13-10. What would you check in the power supply?
6. A set has hum in the sound but not in the picture. The power supply is like Fig. 13-8. What would you check? How would you do it?
7. If the set in Question 6 had good sound, but the picture looked like Fig. 13-11, what would you do? How would you do it?
8. What if the set in Question 6 had hum in both the video and the sound?

CHAPTER **14**

Introduction to Color Television

Almost everyone has tried painting with water colors at some time or another. It does not take much experience with water colors to find out that you should be careful not to get them mixed together, because if the colors are mixed they make other colors. For example, if you have yellow paint on the brush and you stick it into the blue paint, you get a shade of green. Fig. 14-1 shows yellow, blue, and orange mixes.

14-1 Colors and Color Mixing

Mixing colors on a tv screen is not exactly like mixing paints, because you are dealing with transmitted light instead of reflected light. So, if you mix yellow and blue on a tv screen, you do not get green; instead, you get white. And, in fact, you can get yellow by mixing red and green. So white in a tv picture can be produced by using red, green, and blue.

Fig. 14-1. Mixing yellow, blue, and orange water colors.

This is just what is done in a tv set. Using red, green, and blue gives not only these three colors but also white and many other mixtures of red, green, and blue. Fig. 14-2 shows a *color triangle* with red, green, and blue at the corners. This triangle shows what will happen if various combinations of the corner colors are mixed. For example, look at the side between red and green. Starting at red and moving toward green, you see red, orange, yellow, and several shades of green. These are the various mixtures of red and green light.

Fig. 14-2. A color triangle.

If all three colors (red, green and blue) are mixed, white is the result. This is shown in the middle of the triangle. Then, if a little more red is added to the white, you get pink. Colors without any white in them are called *saturated* colors, and colors with some white in them are called *desaturated* colors. Pink, then, is called desaturated red.

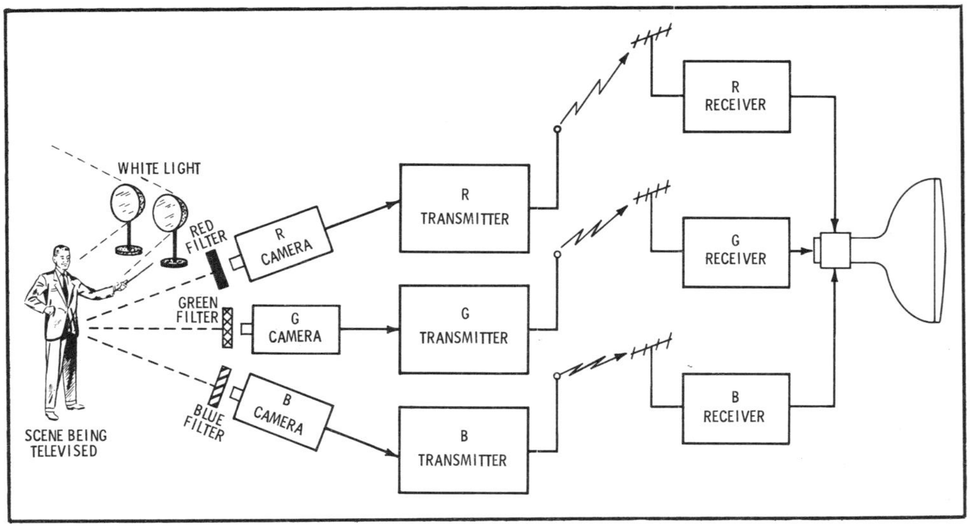

Fig. 14-3. A possible color-television system.

Take pink and add a little more green; the result will be a light orange or yellow. In this way, you can produce a great variety of different colors using only red, green, and blue to start with. Any color, within limitations, can be produced by mixing three colors.

14-2 The Color Television System

As you will recall from Chapter 5, a color-tv picture tube has three electron guns: one for red, one for green, and one for blue. These three guns are aimed so that each one lights up only those phosphor dots having its color. That is, the blue gun only lights up the blue phosphor dots on the screen, the green gun lights up only green dots, and the red gun lights up only red dots.

With this kind of tube, you could build a color tv system using three black-and-white systems. If this were done, you would have three cameras, three transmitters, three receivers, and one picture tube. The system would look something like Fig. 14-3. In such a system, each camera would have filters that allowed it to "see" only one color. The signal from each camera would activate one electron gun in the picture tube.

While such a system could work, it would be very expensive and would be difficult to synchronize properly. Moreover, receivers for such a system could not receive ordinary black-and-white broadcasts. Since color sets must receive black-and-white telecasts, the system shown in Fig. 14-3 is not satisfactory.

The system that is used is thoroughly *compatible* with black-and-white tv. That is, a black-and-white set can receive a color broadcast and show it in black-and-white. Similarly, a color set can receive a black-and-white broadcast and show it

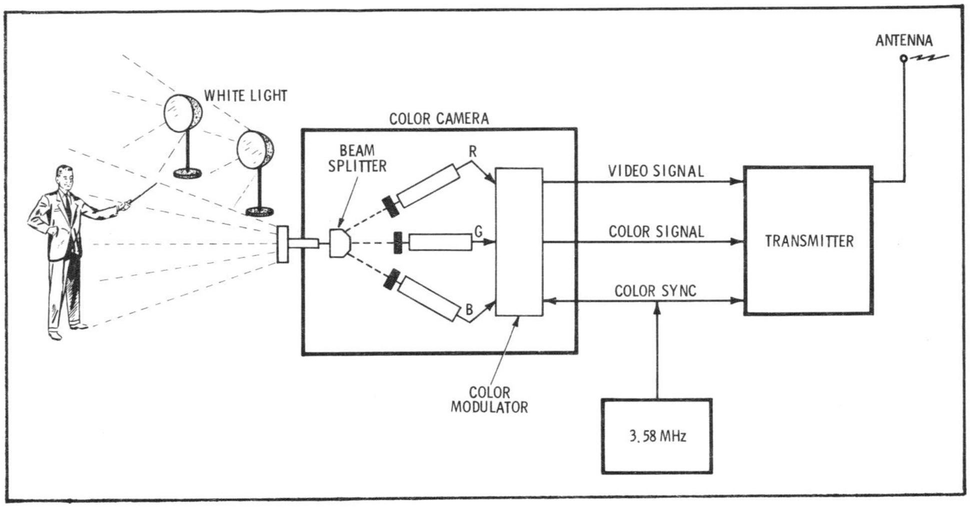

Fig. 14-4. Diagram of color-tv transmitting system.

in black-and-white, and, of course, a color set can receive a color telecast and show it in color. Fig. 14-4 shows a block diagram of the transmission system that is used.

Inside the color camera, the light from the televised scene is divided into its three basic color components so that red, green, and blue signals are obtained. These three signals are sent to a color modulator circuit. This circuit does two things: It combines the three color signals into a

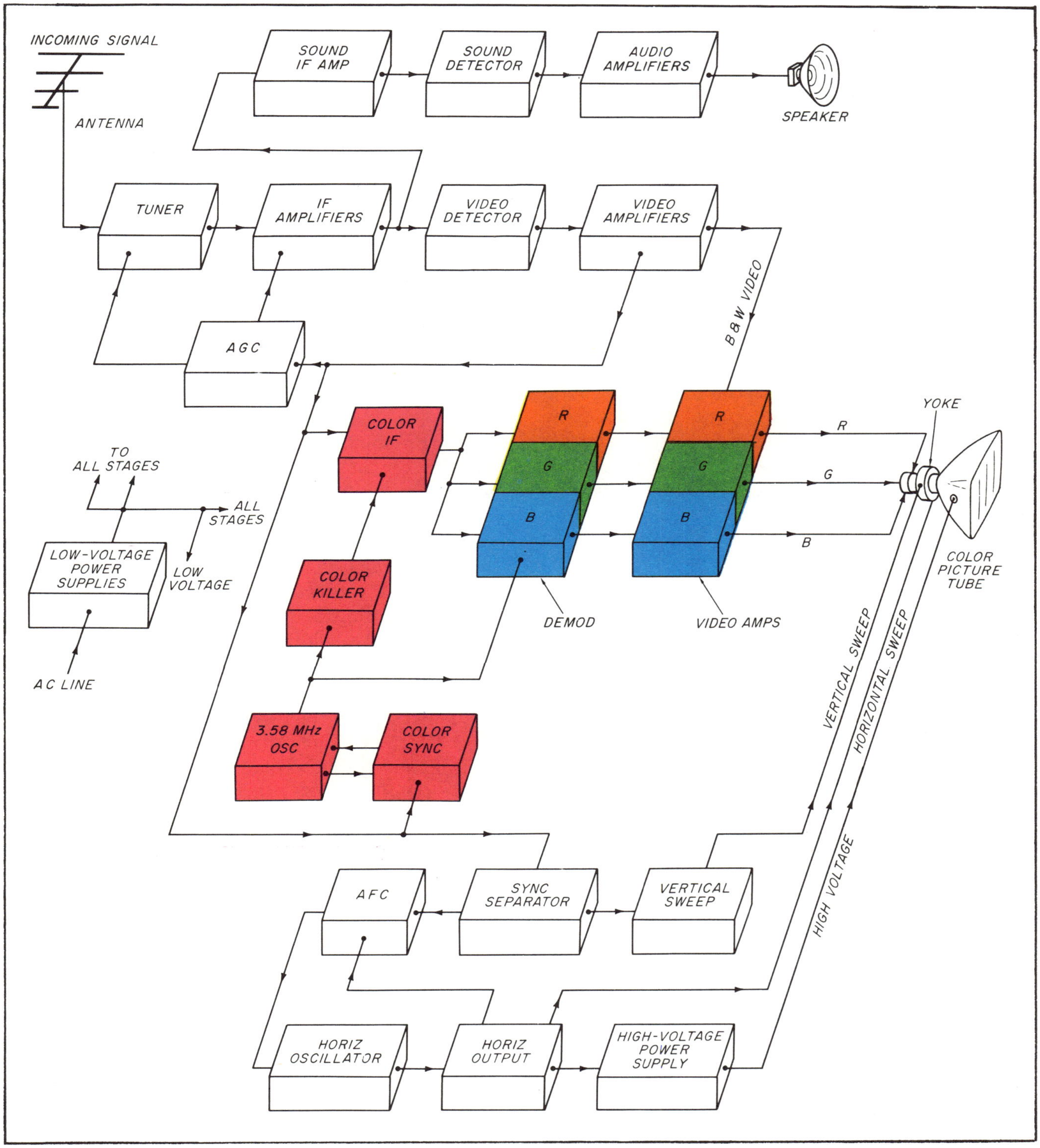

Fig. 14-5. Block diagram of a color receiver.

color signal, and it also produces a separate color-sync signal. The video signal is the same as a black-and-white video signal. The color signal is *phase modulated* on a 3.58-MHz color carrier.

Phase modulation is a process in which the *phase* of the 3.58-MHz carrier is shifted to indicate the *color* of the scene. Imagine a pointer with its center in the middle of Fig. 14-2. If straight up is zero degrees, then we can measure rotation of the pointer in degrees, and each phase angle will correspond to a particular color. Basically, this is how phase modulation of the color carrier (3.58 MHz) by the color signal works.

The color-sync signal is a sample of the unmodulated 3.58-MHz carrier. This signal, the video signal, and the modulated color signal are sent out by the transmitter along with the horizontal- and vertical-sync pulses and the sound. A black-and-white receiver does not respond at all to the 3.58-MHz color signals but does reproduce the black-and-white video signal.

A color set uses the 3.58-MHz sync signal to do two things. First, the sync signal turns on the color section of the set. Then it is used as a reference to *demodulate* the color signals. If a color set does not receive a color-sync signal, the color section stays off, and the set shows a black-and-white picture. This is how a color set handles a black-and-white broadcast.

14-3 The Block Diagram of a Color Receiver

A color receiver must do all of the things a black-and-white set does. And, it must also produce colored pictures. Fig. 14-5 shows the block diagram of a color television receiver. Notice that this block diagram is the same as one for a black-and-white set except that it has six additional blocks. These six blocks (color sync, 3.58-MHz oscillator, color killer, color if, color demodulators, and color video amplifiers) are called the color, or *chroma,* section of the set.

If a black-and-white signal is being received, there is no color-sync signal. With no color-sync signal, the 3.58-MHz oscillator is cut off. This causes the color killer to cut off the color if amplifiers. The color video amplifiers are preset to produce a black-and-white picture.

When a color picture is received, the color-sync signal triggers the 3.58-MHz oscillator and synchronizes it with the station's sync signal. The 3.58-MHz oscillator signal is applied to the color killer and to the color demodulators. The 3.58-MHz signal causes the color killer to turn on the color if amplifiers. These amplifiers may be called bandpass, chroma-bandpass, or color if stages. The color if amplifiers amplify the color signal and feed it to the color demodulators. In the demodulators, the color signal is compared to the 3.58-MHz oscillator signal, and the original red, green, and blue video signals are recovered. The three color components are fed to the video amplifiers. In the color video amplifiers, the three color signals are combined with the black-and-white signal and fed to the picture tube. The picture tube then shows the color picture being televised. In many sets, the color and black-and-white signals are combined at the picture tube instead of in the video amplifier.

Receivers that use the *Trinitron* single-gun picture tube have a somewhat different chroma section. While this particular system will not be discussed here, you should look it up in an electronics magazine and read about it.

14-4 Color-Section Circuitry

For practical purposes, it is convenient to group several of the color stages together. Fig. 14-6 shows the circuitry of these stages in a solid-state color receiver. In this circuit, transistor Q7S is the color-sync stage. It is cut off all of the time except when the color-sync signal is present. The color-sync signal (or *color burst*) always occurs just after a horizontal-sync pulse. So, in this circuit, the horizontal-deflection pulse is applied through Q8F and Q10S to turn on the color-sync stage at the correct time. When the color-sync stage (Q7S) is turned on, the color-sync signal is applied to the 3.58-MHz crystal, causing it to oscillate. The output of the crystal oscillator is applied to the color killer (Q5S and Q6S) through the acc (automatic chroma control) amplifier (Q4S). Then, the color killer controls the bias of the second color if stage (Q2S). The action of the acc (automatic chroma control) and color killer are just like agc in that they control the gain of the color if amplifier. The output of the crystal oscillator is also fed through Q8S, Q11S, and Q12S to the color demodulators.

Color-section circuitry may also be constructed using vacuum tubes. Fig. 14-7 shows such a case. In this circuit, the color-sync signal is applied to the burst amplifier (V18), which is gated by the horizontal-flyback pulse. When the burst amplifier is turned on, the color-sync signal is fed to the chroma-sync phase detector (V19). The 3.58-MHz

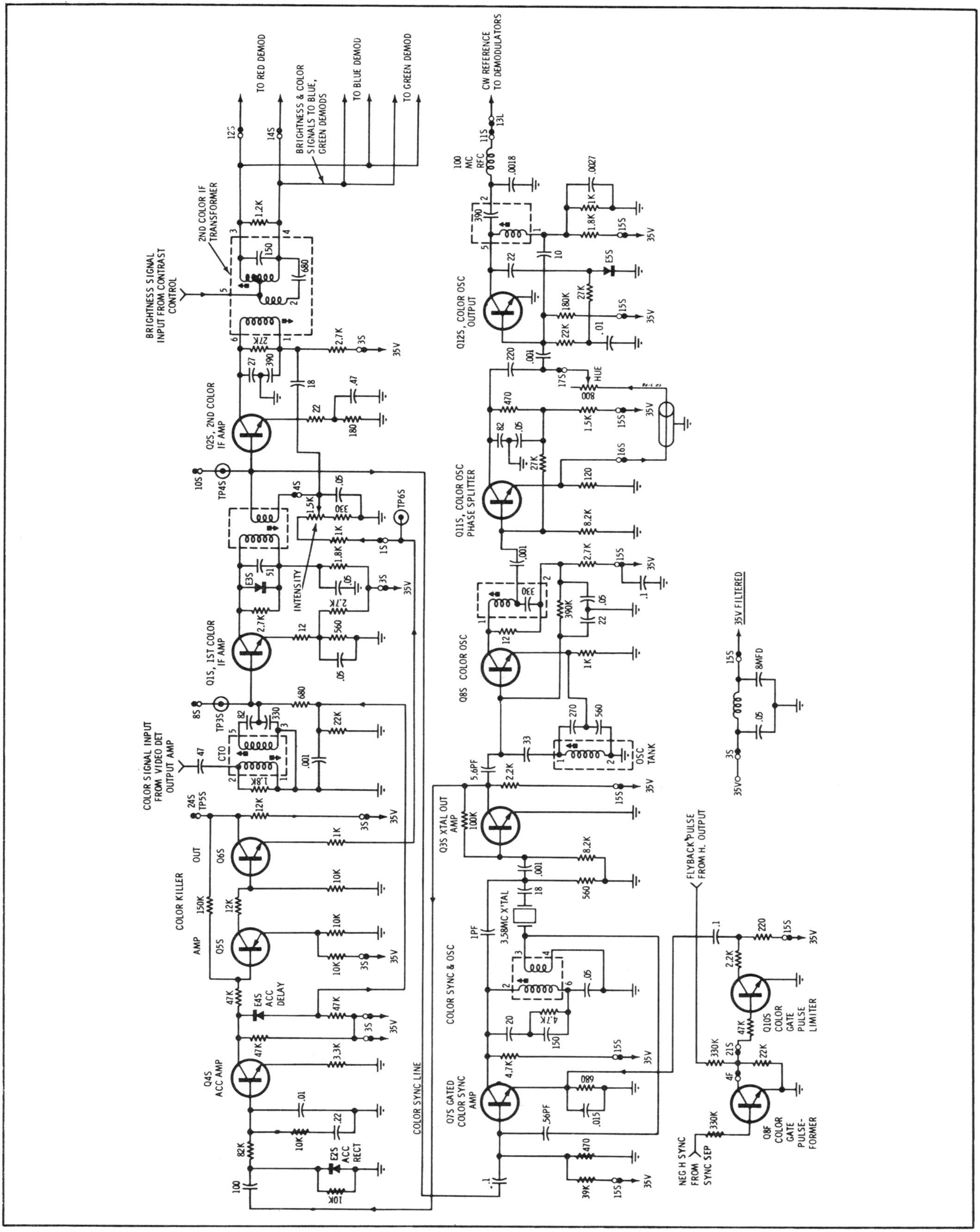

Courtesy Motorola Communications & Electronics, Inc.

Fig. 14-6. Solid-state color-sync circuitry.

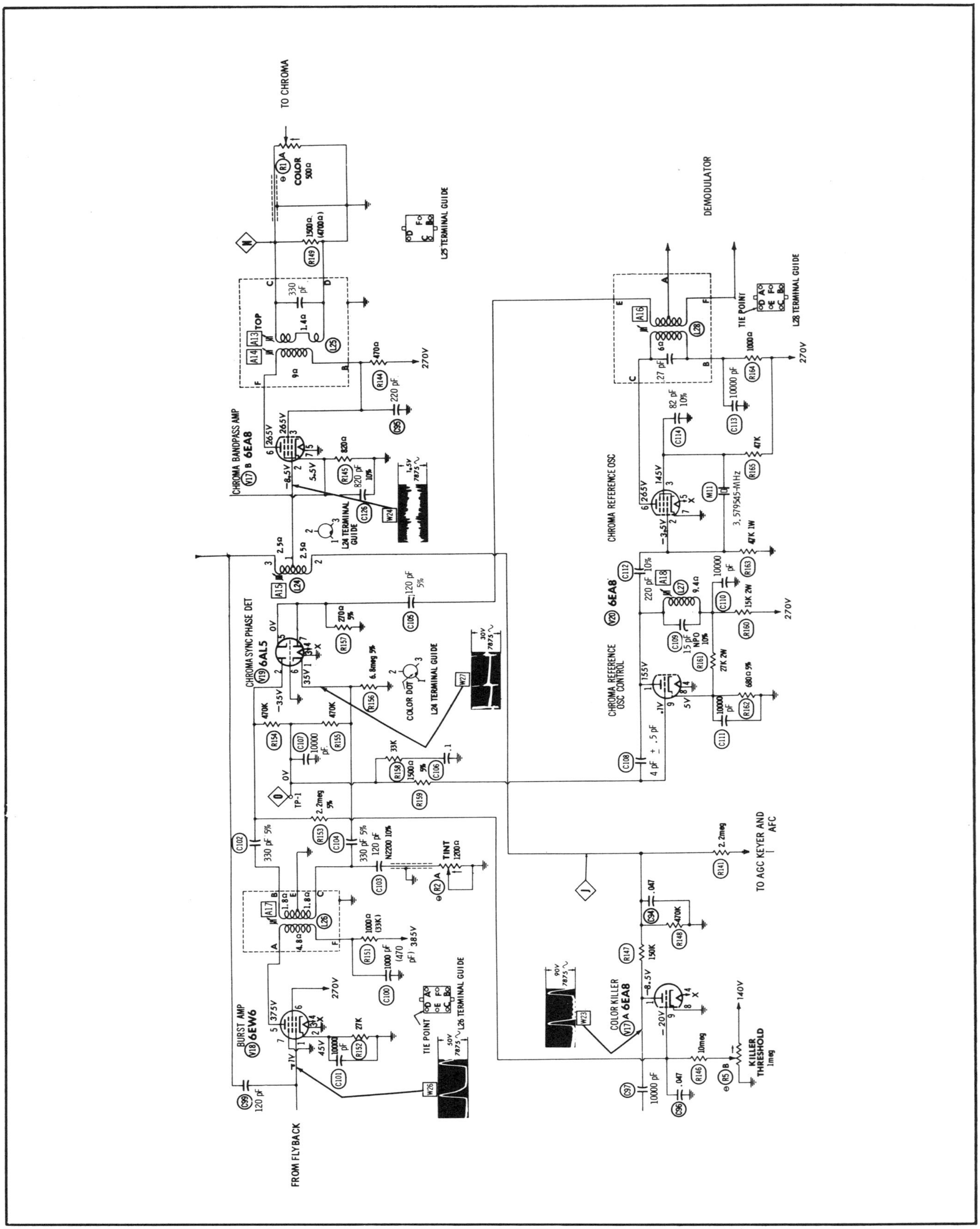

Fig. 14-7. Tube-type color circuitry.

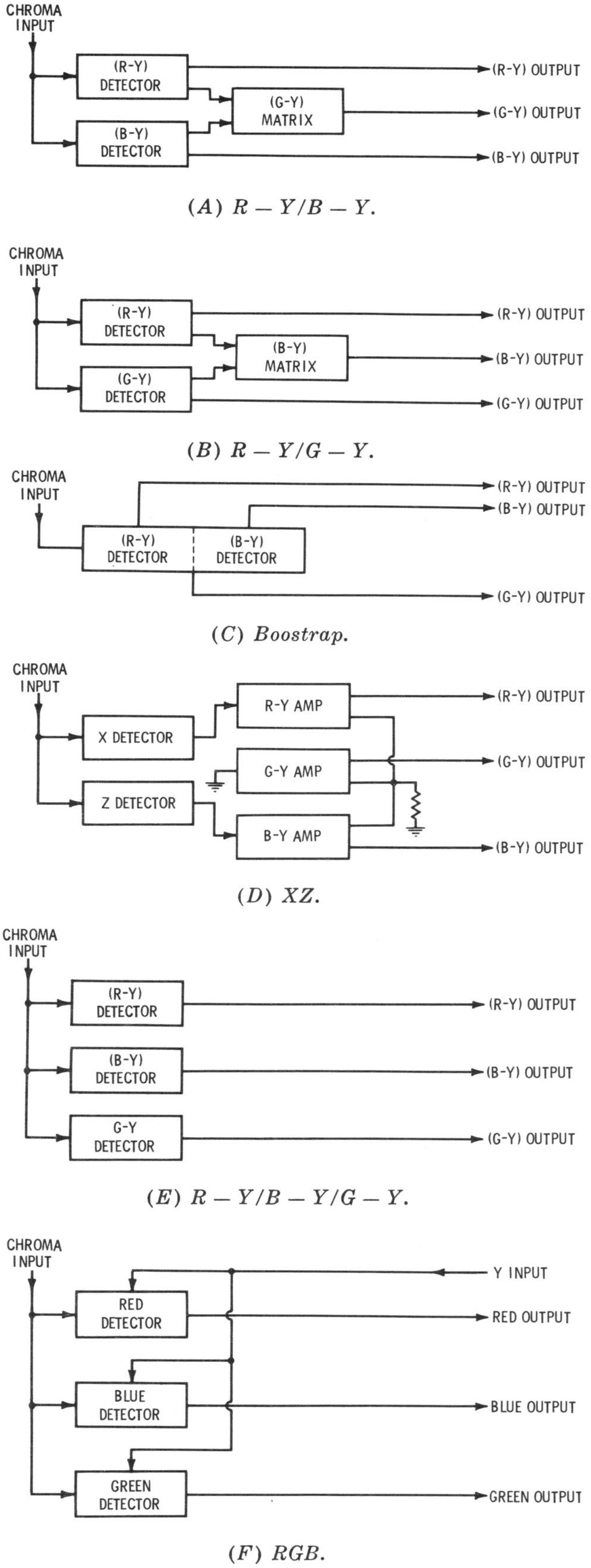

(*A*) $R - Y/B - Y$.

(*B*) $R - Y/G - Y$.

(*C*) *Boostrap.*

(*D*) *XZ.*

(*E*) $R - Y/B - Y/G - Y$.

(*F*) *RGB.*

Fig. 14-8. Color-demodulator systems.

oscillator (V20) is synchronized with the burst by the action of the chroma-sync phase detector and the oscillator control circuit. The output of the oscillator is to both the phase detector and the demodulators. The color killer (V17A) also is controlled from the output of the burst amplifier. The color killer in turn controls the chroma-bandpass amplifier (V17B), which is the first color if stage.

14-5 Color Demodulators

There are several different types of color-demodulator circuits. Fig. 14-8 shows several of the most common types in block-diagram form. In these block diagrams, the Y signal is the luminance, or black-and-white, video signal. The R, G, and B signals are the red, green, and blue signals, respectively. The signals labeled R − Y, G − Y, and B − Y are what remains of the red, green, and blue signals after the black-and-white part has been removed. In other words, they are the signals that must be combined with the black-and-white signal to get red, green, and blue signals.

Since the color signals are phase-modulated on the 3.58-MHz color carrier, all of the demodulators use phase-detector circuits. In these circuits, the color signal is compared with the 3.58-MHz oscillator signal, and an output proportional to the phase difference is produced. Fig. 14-9 shows a typical set of color demodulators. Notice that the three demodulator circuits are *almost* identical, the only difference being the way the 3.58-MHz reference signal is coupled into the individual detector. In the phase-modulation system used, the only differences between the three signals are their amplitudes and the phase angles between them and the 3.58-MHz reference.

Color demodulators, like almost all other electronic circuits, can also be made using vacuum tubes. Fig. 14-10 shows a vacuum-tube demodulator circuit used to produce R − Y, G − Y, and B − Y video signals. In systems using R − Y, G − Y, and B − Y signals, the Y signal is often added at the picture tube as shown in Fig. 14-11. Of course, if you add Y to R − Y, G − Y and B − Y, the result is simply R, G, and B.

Since all demodulators do basically the same thing and their circuits are relatively simple, you might expect that they could be made in integrated-circuit form. This is indeed the case. Fig. 14-12 shows the equivalent circuit of one integrated color-demodulator "chip." There is little

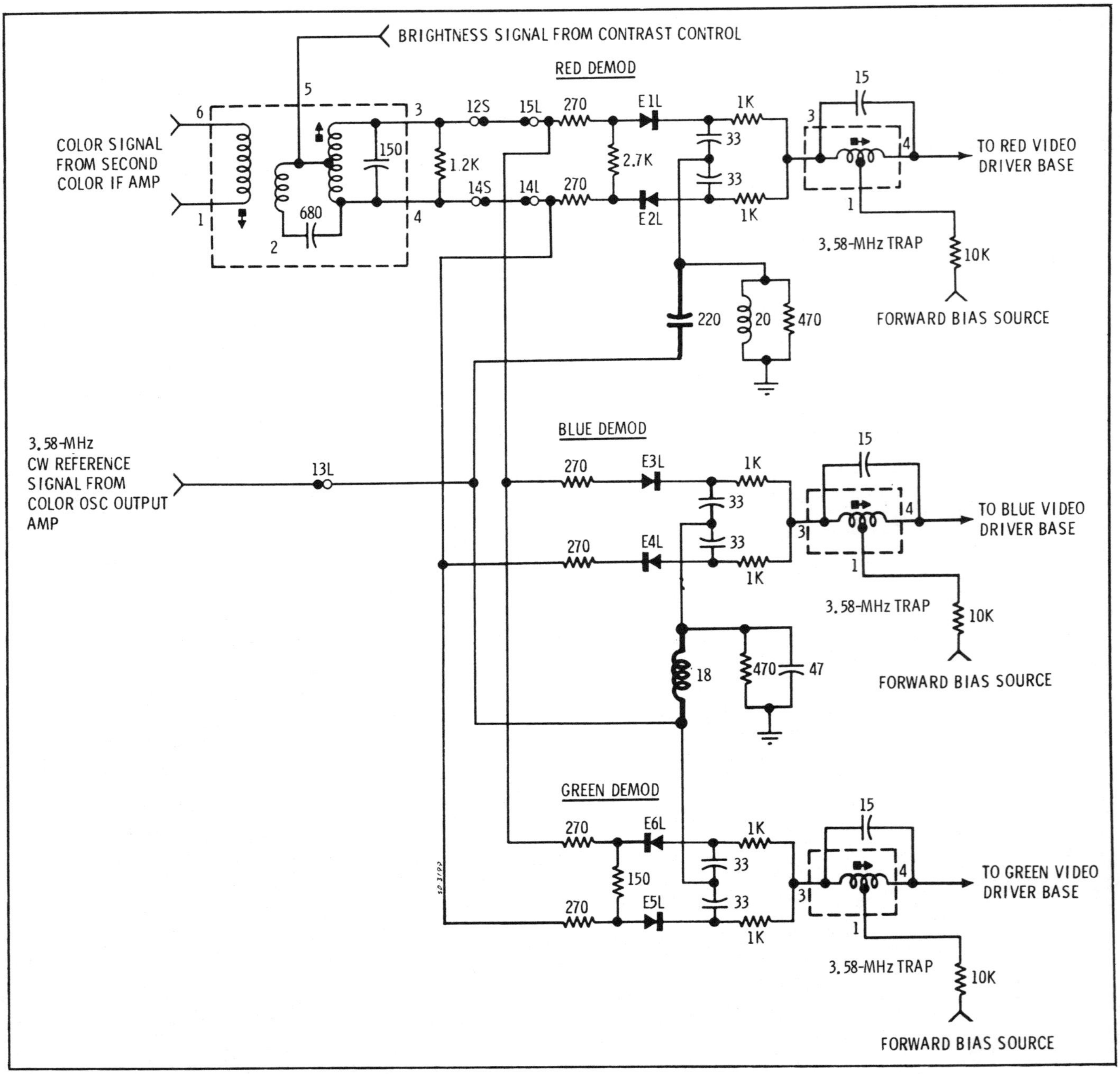

Fig. 14-9. Typical color-demodulator section.

doubt that more and more ICs will be used in television receivers.

14-6 Color Video-Amplifier Circuits

Just as there are a variety of color-demodulator systems, there are also several video-amplifier arrangements. Fig. 14-13 shows one type of amplifier that is used with an RGB demodulator like the one in Fig. 14-9. Notice that in the RGB demodulator, the black-and-white signal (Y) is combined with R − Y, G − Y, and B− Y in the demodulator. There is, therefore, no need to add the Y signal at the picture tube or in the video amplifiers. When this system is used, the amplifiers are referred to simply as red driver, blue output, etc.

When one of the R − Y, G − Y, B − Y systems is used, the video amplifiers are usually called *color-difference* amplifiers. Fig. 14-14 shows a typical circuit of this type.

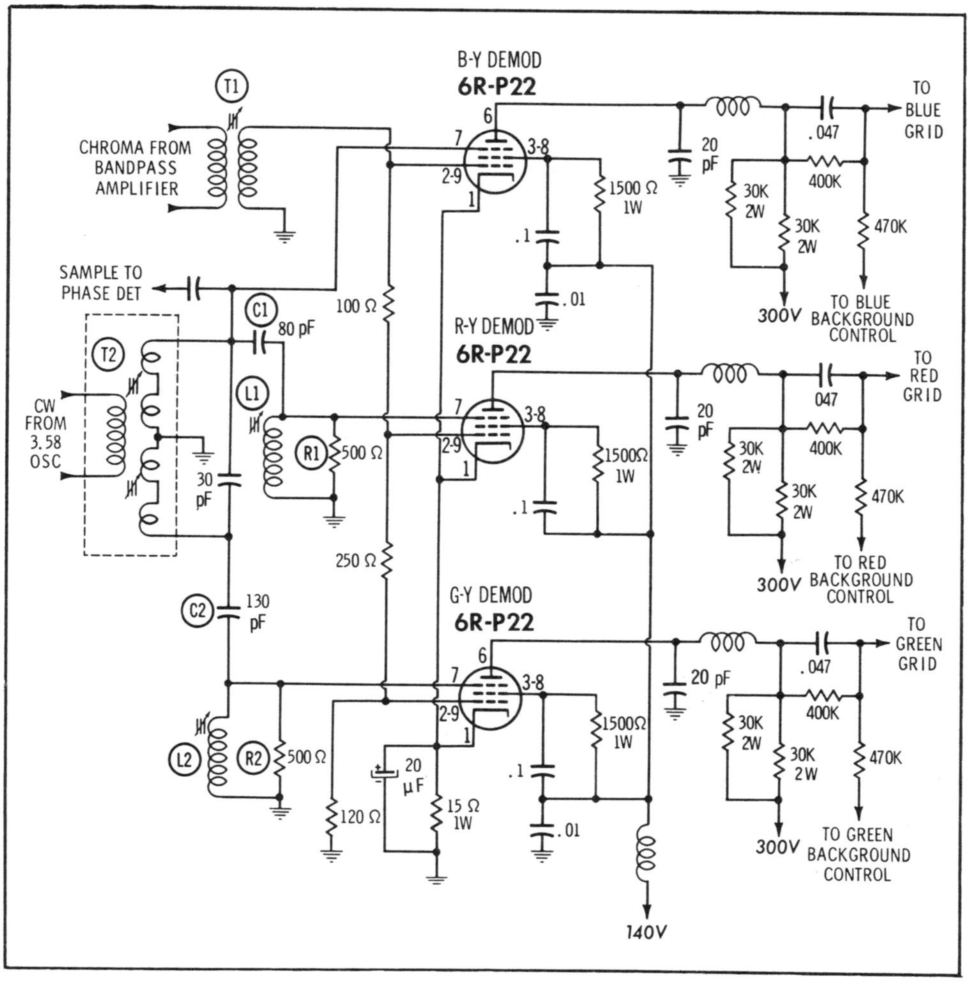

Fig. 14-10. Tube-type color demodulator circuit.

14-7 Review Questions

1. What are the three basic colors used in a color tv set?
2. What colors are at the corners of a color triangle?

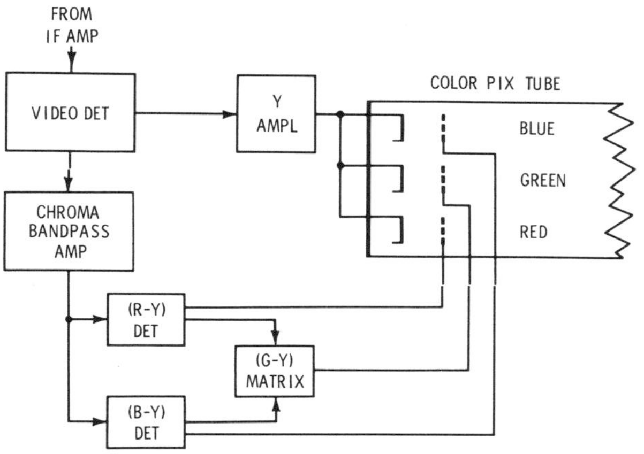

Fig. 14-11. Adding Y to R − Y, G − Y, and B − Y.

3. What color do you get on a tv screen if you mix equal parts of:

   A. red and green?
   B. red and blue?
   C. blue and green?
   D. red, blue, and green?

4. What is meant by color saturation?
5. Name one desaturated color.
6. How many electron guns does a color-tv picture tube usually have?
7. What is meant by *compatible* color tv?
8. What is the frequency of the color-sync signal?
9. What are the names of the stages usually found in a color section?
10. In what ways are a black-and-white set and a color set alike? In what ways are they different?
11. What does a color killer do?

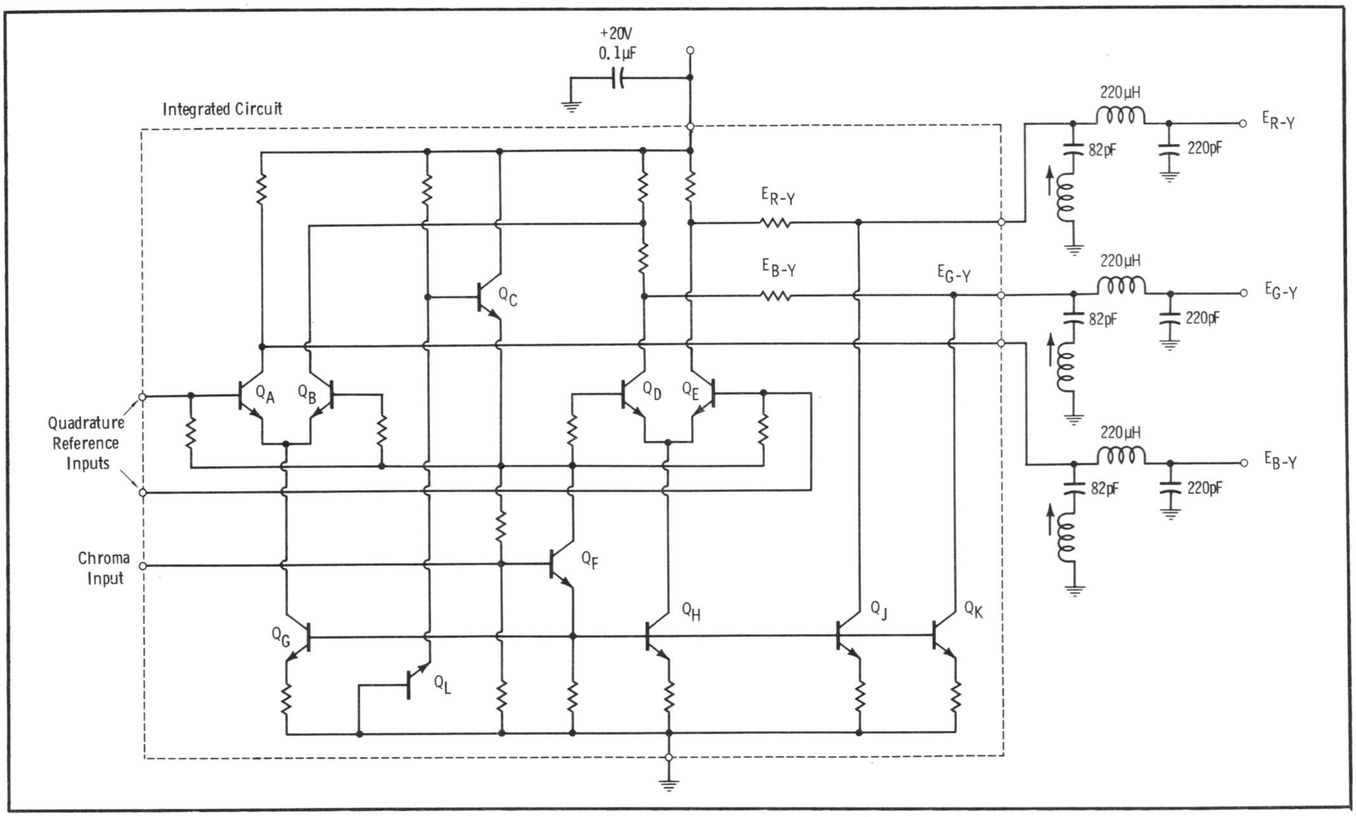

Fig. 14-12. A chroma-demodulator integrated circuit.

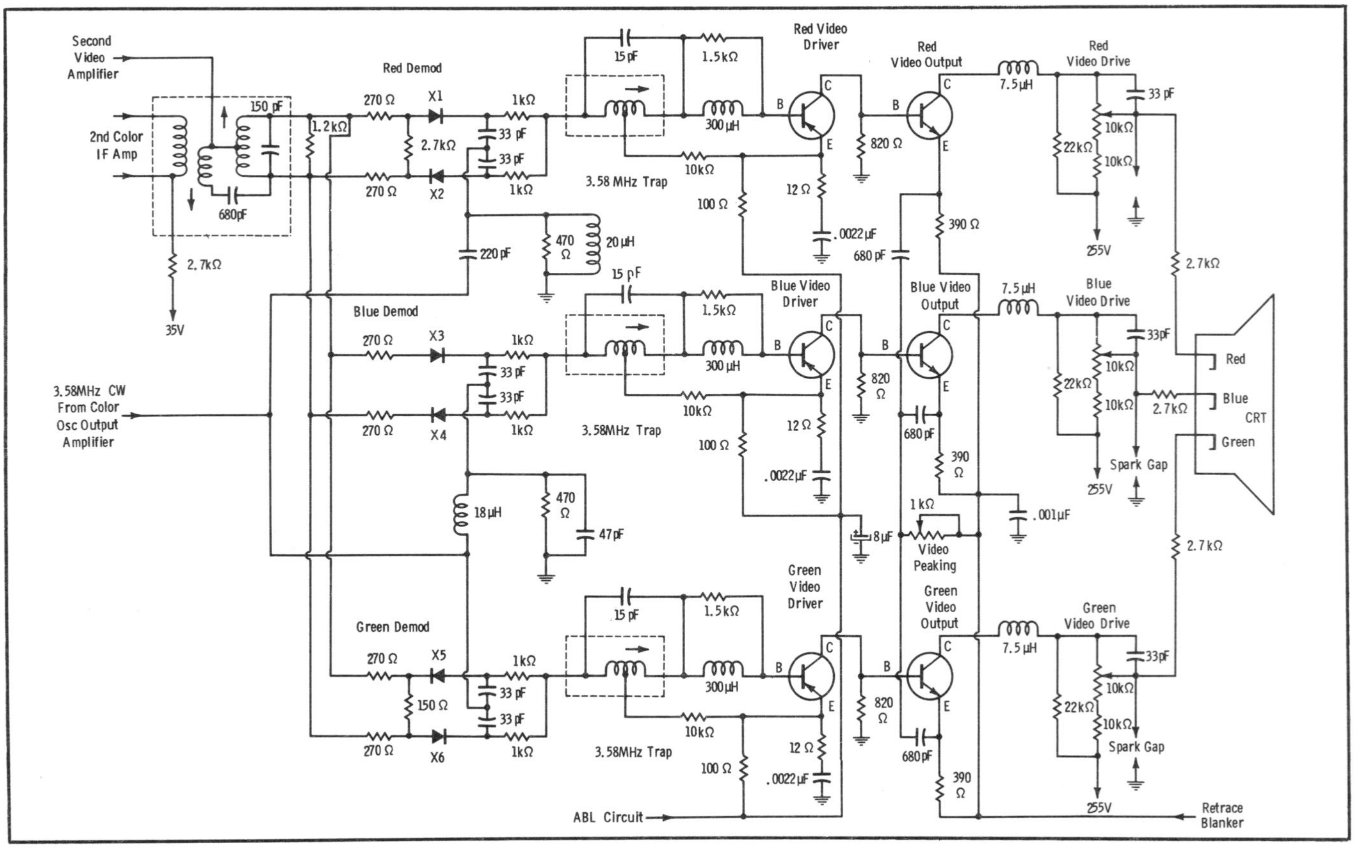

Fig. 14-13. Color video amplifiers.

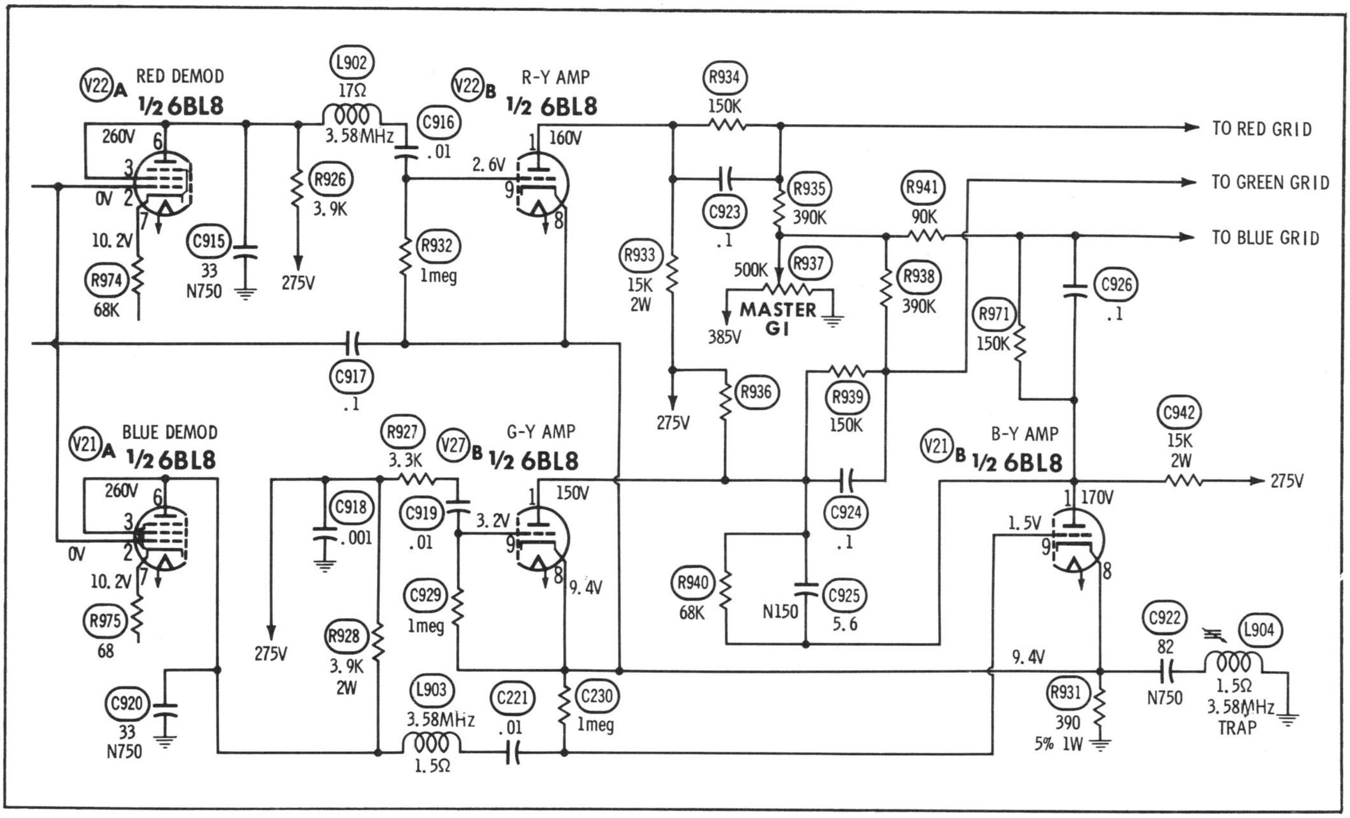

Fig. 14-14. Color-difference amplifiers.

12. What is another name for the first color if stage?
13. What does acc mean? What is acc for?
14. What kind of modulation is used with the color video signals?
15. Draw block diagrams of four types of color-demodulator circuits.
16. What is a luminance signal?
17. Tell what each of the following signals is:

    A. Y.
    B. B.
    C. G − Y.

18. What does the 3.58-MHz oscillator in a color set do?
19. What cuts off the color when a black-and-white telecast is being viewed?
20. What is a color-difference amplifier?

CHAPTER **15**

Receiver Setup Adjustments

Many of the problems that are seen in television pictures can be traced to improperly adjusted controls. In this chapter, the various controls and their effect on set performance will be considered. In particular, we will be concerned with the proper adjustment of the controls that affect the color picture. Adjustment of single-gun *Trinitron* sets will not be discussed. However, you may wish to investigate this area on your own.

15-1 Black-and-White Adjustments

The first group of controls that one encounters in a tv set are the "customer" controls. These include:

1. The on-off switch
2. The volume and tone controls
3. The channel selector and fine-tuning control
4. The contrast and brightness controls
5. The vertical- and horizontal-hold controls

You should already know how to set each of these controls for normal viewing.

Inside the set on the chassis there are many other adjustments and controls. Some of them are controls that should be adjusted as needed during installation or service. These controls can be called "service" controls. In this category are such adjustments as:

1. Vertical height and linearity
2. Agc
3. Focus

Vertical height and linearity should be set with a test pattern, crosshatch pattern, or white-dot pattern. The objective is to get a full-height picture with good linearity. Since the height and vertical-linearity controls interact with each other, it is necessary to adjust both alternately several times to get good results. If none of the test patterns named above are available, approximate height and linearity settings can be made with any picture.

Manufacturer's service data and Sams PHOTOFACT® usually include a procedure for setting the agc level. If this information is not available, the agc can be set approximately by tuning to a strong station. Then advance the agc control until the picture becomes slightly bent and contrasty. Finally, back off the control a little until no bending is observed. This sets the control so that the receiver will not overload on the normal strong-signal level. Check the picture on other stations to be sure there is no bending.

The focus control can usually best be set with the aid of a test pattern. Adjust the control for the sharpest focus when you view the screen from a distance of about five times the height of the screen.

There are many other controls located on the chassis of a tv set. Some of these adjustments should be attempted *only* by expert technicians who have had advanced training. For that reason, we shall call them "advanced" adjustments. They include:

1. Tuner alignment
2. If alignment
3. Horizontal frequency and drive
4. Sound traps and sound alignment

Since many of these adjustments require special training and equipment, we will not discuss them.

15-2 Color Receiver Setup Controls

Color sets have all the controls of a black-and-white set and several others that are needed for color. Among the additional "customer" controls you may find:

1. Tint, hue, and color intensity
2. Sharpness or peaking
3. Aft switch

You should consult service information if you do not know how to adjust each of these controls for normal viewing. Usually, the color controls should be set in the middle of their ranges while other setup adjustments are made.

In addition to "customer" controls, a color set has a group of "service" adjustments not found in a black-and-white set. Among them are:

1. High-voltage adjustment
2. Color purity adjustment
3. Static convergence adjustments
4. Dynamic convergence adjustments
5. Screen color controls
6. Color-killer threshold control
7. A "service" switch

These will be dealt with in some detail later. For the moment, just observe that these adjustments *must* be made correctly if the color set is to produce a good color or black-and-white picture.

There are also several "factory" adjustments in a color set that are not found in black-and-white receivers. Among these are:

1. Color if alignment
2. 3.58-MHz traps
3. Color-demodulator alignment

Normally, these adjustments should be made only by specially trained expert technicians.

15-3 Preliminary Color Setup Procedures

Before you try to make the actual color setup, there are several things that must be done.

Check and adjust the high-voltage setting. *Follow the manufacturer's instructions exactly.* If you do not have access to these instructions, *DO NOT ADJUST THE HIGH VOLTAGE.* Improperly adjusted high voltage can cause a poor quality picture and shortened picture-tube life and may cause dangerous X-ray emission.

Check black-and-white "customer" and "service" adjustments. They should all be set correctly for normal viewing.

Magnetic fields can cause discoloring in a tv picture. So, it is necessary to *degauss* the set before doing color setup. The effect of magnetization on a tv picture can be seen in Fig. 15-1. Notice that both the black-and-white picture (Fig. 15-1A) and the color picture (Fig. 15-1B) are affected. Fig. 15-2 shows a degaussing coil that can be used to remove unwanted magnetic effects.

The procedure to follow in degaussing a set is to plug the degaussing coil into a 120-volt ac outlet. Then, holding the coil several feet (4 to 6) from the set, turn it on. Now, slowly bring the coil up to the set. Move it slowly over the top, sides, bottom and front of the set. The coil should be held flat and close to (but not touching) the picture-tube face while this is done. Finally move the coil away several feet (6 to 8) and turn it off.

(A) Test pattern.

(B) Bar pattern.

Fig. 15-1. Magnetic discoloration in tv pictures.

Fig. 15-2. Degaussing coil.

While the set is being degaussed, a wild, almost psychedelic pattern like Fig. 15-3 can be seen on the screen. This is normal.

Do not hold the degaussing coil near the back of the set. This could damage some of the magnetic components located on the neck of the picture tube.

Summarizing to this point, recall that there are three preliminary setup operations. They are:

1. Adjust the high voltage according to manufacturer's recommendations.
2. Adjust the black-and-white controls for normal viewing.
3. Degauss the set.

When these things have been done, continue with the setup of the color receiver.

15-4 Adjusting Picture-Tube Purity

When a color picture tube is properly adjusted, the electrons from the red gun strike only red phosphor dots. Similarly, electrons from the green and blue guns strike only green and blue dots, respectively.

When this does not happen, it is said that the screen has poor purity. The purity of a tv screen is adjusted by aiming all three electron beams. This is done by the positioning of two circular magnets on the neck of the picture tube. These purity rings are indicated in Fig. 15-4. (Purity and convergence of an in-line system will be given in Paragraph 15-6. The following discussion applies to delta-system sets only.)

To make the purity adjustment, all the video to the picture tube must be cut off. In most sets, this can be done easily by means of the *service switch* on the chassis. Just set the service switch to the *raster* position. If the set does not have a service

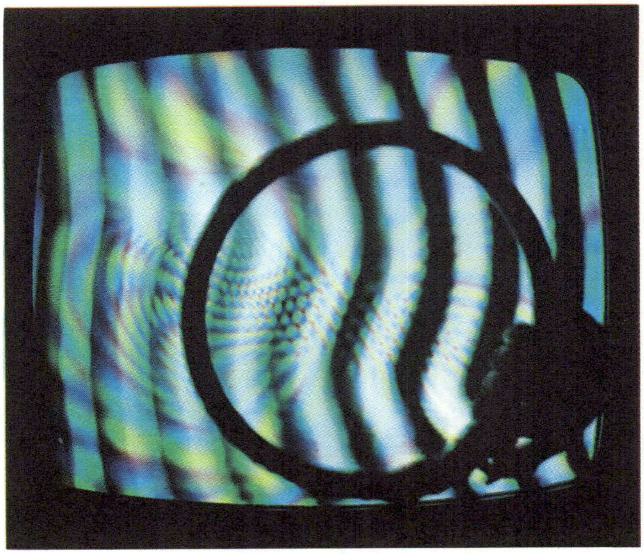

Fig. 15-3. The degaussing pattern.

switch, then purity adjustments are usually made with the set tuned to an unused channel and the antenna disconnected. Setting the contrast control to minimum also helps.

Before the purity adjustment is made, it is necessary that the static convergence be at least approximately correct. In most cases, this presents no problem. However, occasionally the convergence will be so bad that you will have to go through the procedure in Paragraph 15-6 before making the purity adjustment.

With the video off, find the red-, green-, and blue-screen controls on the chassis. Turn off the

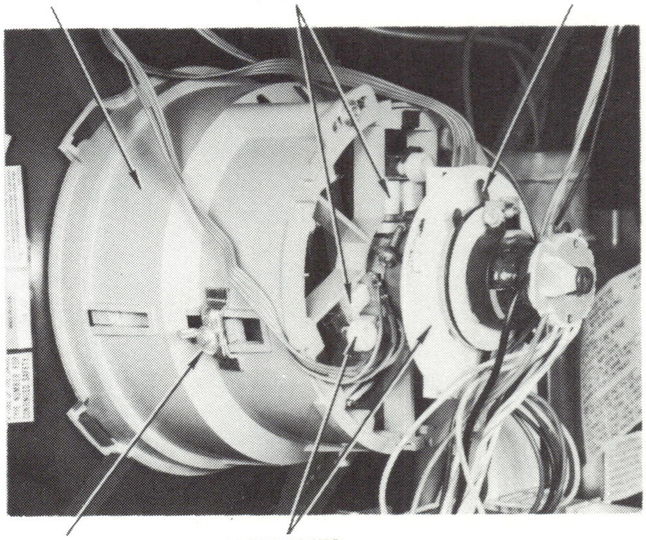

Fig. 15-4. Yoke components.

Fig. 15-5. Poor red-screen purity.

green and blue screens. This should give you a full red screen. If the purity is poor, the screen will be discolored as shown in Fig. 15-5.

To correct the purity, loosen the thumbscrews on the sides of the yoke. Then carefully slide the yoke backward as far as it will go. Be very careful not to disturb any of the other components on the neck of the tube.

The screen will now be severely discolored as shown in Fig. 15-6. Watch this "purity pattern" carefully and adjust the two purity rings on the neck of the tube. A mirror placed in front of the set will be very helpful in doing this. Set the rings so that the red region of the purity pattern is *centered* on the screen.

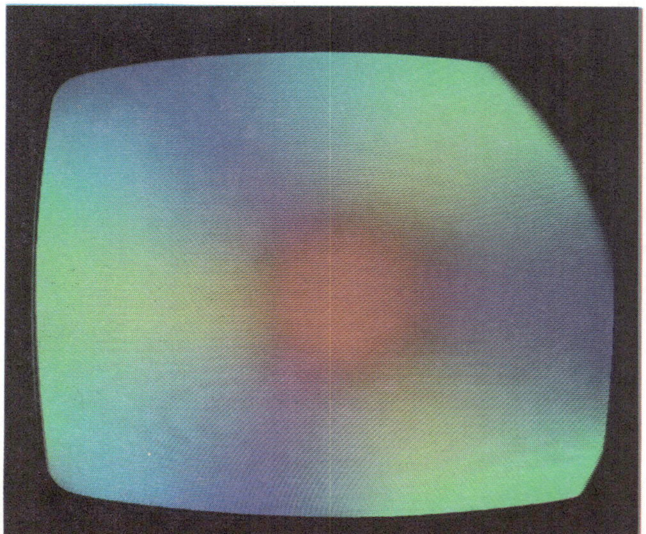

Fig. 15-6. Purity pattern with yoke pulled back.

(*A*) *Pure red screen.*

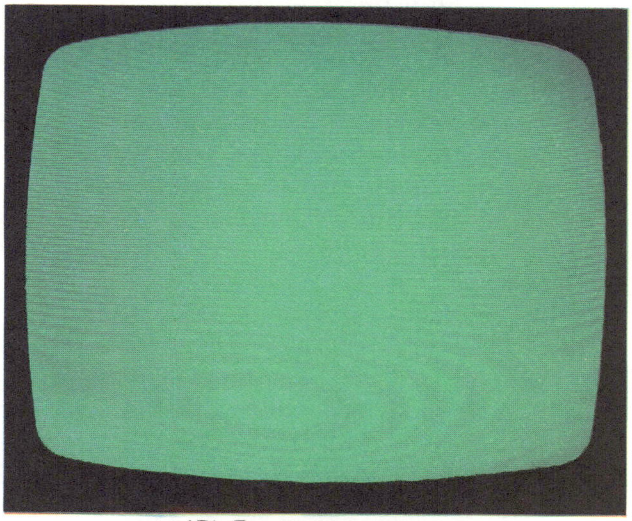

(*B*) *Pure green screen.*

(*C*) *Pure blue screen.*

Fig. 15-7. The three color screens after purity adjustments.

Now carefully slide the yoke forward while observing the screen. Find the yoke position that gives you the best full red screen. There should be no discoloration anywhere on the screen. At this point, tighten the thumbscrews on the yoke.

After the red-screen purity is set, check both the green and blue screens. By turning off the red-screen control and turning on the green, you can see the green-screen purity. Similarly, by turning off the red and green while turning on the blue, you can see the blue-screen purity. All three screens should be pure as shown in Fig. 15-7. If one of the screens is slightly discolored, repeat the whole purity procedure again. When you have satisfactory purity on all three screens, return the service switch to the "normal" position.

15-5 Adjusting the Gray Scale

After the purity setup is complete, it is necessary to reset the red-, green-, and blue-screen controls for normal operation. This process is called setting the *gray scale.* There are two different methods for making gray-scale adjustments. The first and perhaps most often used is for sets with a service switch. The second is for sets without a service switch.

If the set has a service switch, set it to the "service" position. Set the crt bias control to its fully counterclockwise position. Then set the brightness and drive controls to their fully clockwise positions. Turn all three screen controls off. Slowly turn up the red-screen control and watch the screen. At first the screen will remain dark, then as you turn up the red-screen control, a red line will suddenly appear. When this occurs, reduce the screen-control setting until the line just vanishes. Repeat this process for the green- and blue-screen controls.

When all three screen controls have been set, return the service switch to the "normal" position. The screen should now be its normal white color. If not, touch up the screen controls very slightly as required to produce a white screen.

If the set has no service switch, tune to an unused channel. Then disconnect the antenna and turn the contrast all the way down. Turn off the green- and blue-screen controls. Set the red-screen control almost all the way on. The screen should be bright red. Next turn up the green-screen control until the screen is bright yellow. Finally turn up the blue-screen control until the screen is its normal white color.

If you cannot get a satisfactory yellow or white screen, start over and do not set the red quite so high.

At this point it may be impossible to get the screen to be *all* yellow or white. If so, the set may be out of convergence. We will come back and reset the gray scale after we finish the convergence procedures.

15-6 Static Convergence

So far, the red gun is striking only red dots, the green gun is striking only green dots, and the blue gun is striking only blue dots. This condition gives pure color screens. But, the three guns may not be striking the same color triad at the same time. (A triad is one set of red, green, and blue dots.) For instance, the red gun may be striking one color triad while the green is striking the triad above. Similarly the blue gun may be striking a triad to one side or the other.

What you must do now is aim each beam individually so that all three guns strike the same triads at the same time. This is first done in the center of the screen where the deflection system has its least effect. This process of aiming the electron beams at the center of the screen is called *static convergence* of the set.

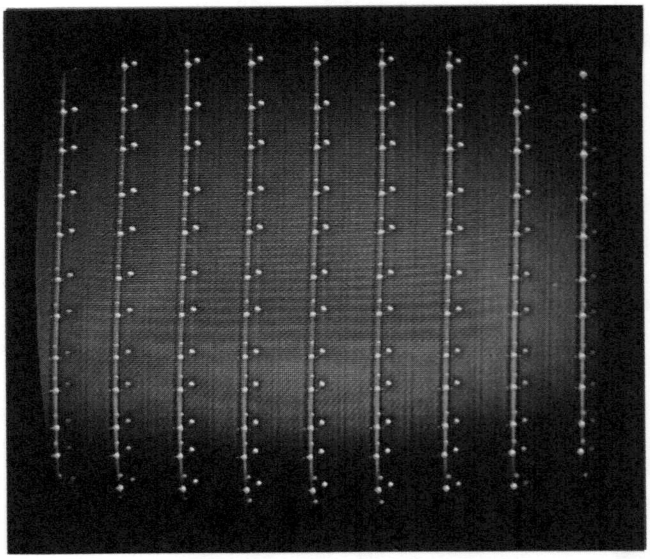

Fig. 15-8. Poor static convergence.

A white-dot generator is usually used in making the static-convergence adjustments. The generator is connected to the antenna terminals, and the set is adjusted for a clear dot pattern. Fig. 15-8 shows a badly misconverged pattern.

In the case of a set with an in-line color tube, the convergence is performed with two sets of ring magnets located on the tube neck. One pair adjusts the blue and green beams *simultaneously in opposite directions.* The other magnet pair adjusts the same beams *in the same direction.* The two rings are adjusted for good overall convergence (white dots) on the screen. Some in-line tubes come with assemblies premounted and adjusted; they should not be readjusted by the technician.

The static-convergence adjustments of a delta-system tube are mounted on the picture-tube neck just behind the yoke. Fig. 15-9 shows the location of these controls. There are four controls all together. They are:

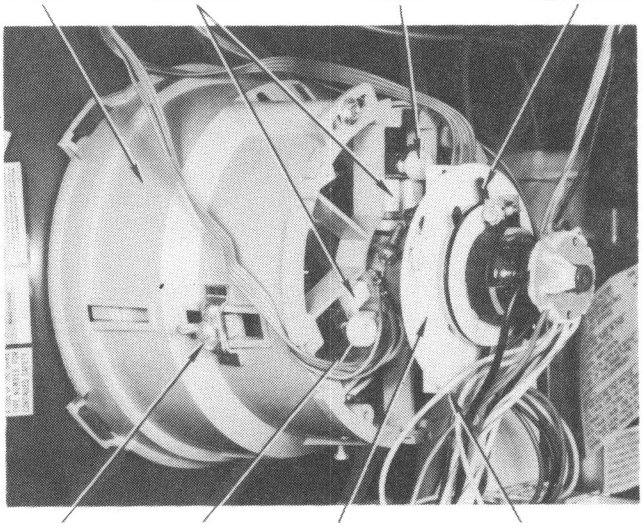

Fig. 15-9. Static-convergence controls.

1. The red-convergence magnet
2. The green-convergence magnet
3. The blue-convergence magnet
4. The blue-lateral magnet

With the dot pattern on the screen, adjust the red- and green-convergence magnets together until the red and green dots merge into yellow dots. Both the red and green controls move the dots diagonally, and they interact with each other. It is not hard to get the red and green dots to form yellow dots as shown in Fig. 15-10. Then adjust the blue-convergence magnet until the yellow and blue dots are side by side as shown in Fig. 15-11. The blue magnet will interact some with the red and green, so the best approach is to move the blue dots part way and then touch up the red and green convergence as needed to get yellow. After this has been done several times, the blue dots will be side-by-side with the yellow dots.

Fig. 15-10. After adjustment of red- and green-convergence magnets.

Now adjust the blue-lateral magnet to move the blue dots into the yellow and form white dots. Again there will be interaction. The best way is to move the blue dots part way, then touch up the red-, green-, and blue-convergence magnets. Repeat the process until white dots are formed in the *center* portion of the screen. Fig. 15-12 shows a screen that has good static convergence.

Fig. 15-11. After adjustment of blue-convergence magnet.

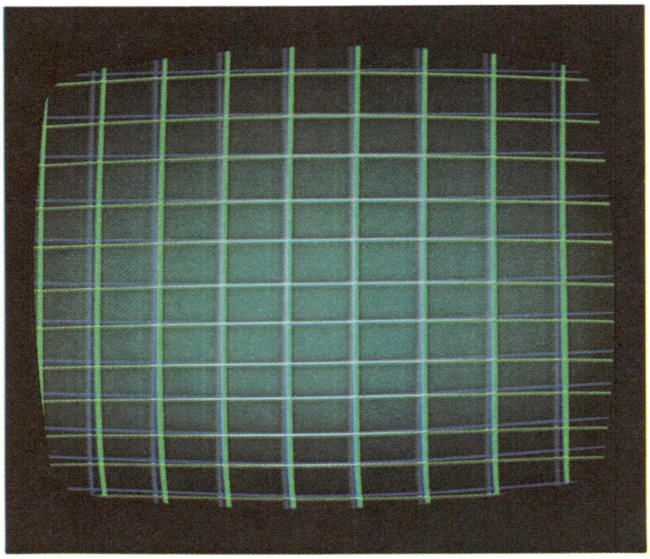

Fig. 15-12. Good static convergence.

15-7 Dynamic Convergence

After the set is statically converged, the dots may still not be converged around the edges of the screen. The process used to converge at the edges is called *dynamic convergence.* (One of the advantages of the in-line tube system is that *it does not require dynamic convergence.* When the purity/convergence rings have been properly adjusted, the setup of this type of set is complete.)

Dynamic convergence of a delta-system set is usually carried out with a crosshatch generator. The generator is connected to the antenna terminals, and the set is adjusted for a sharp picture. Fig. 15-13 shows a crosshatch pattern on a set

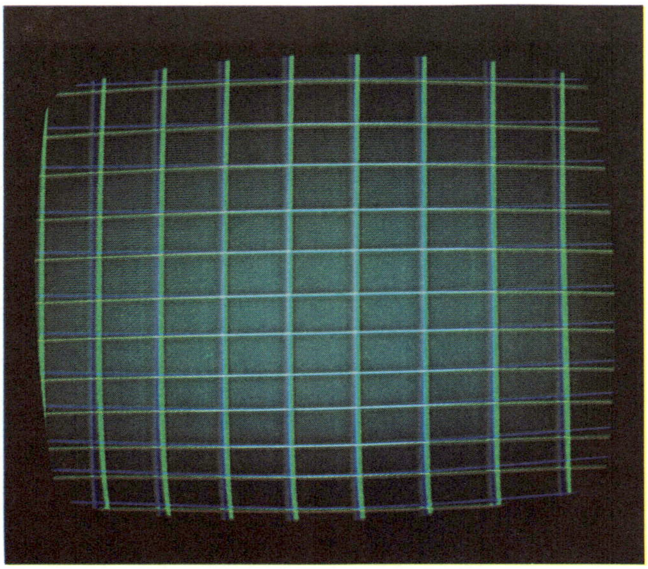

Fig. 15-13. Pattern for dynamic convergence.

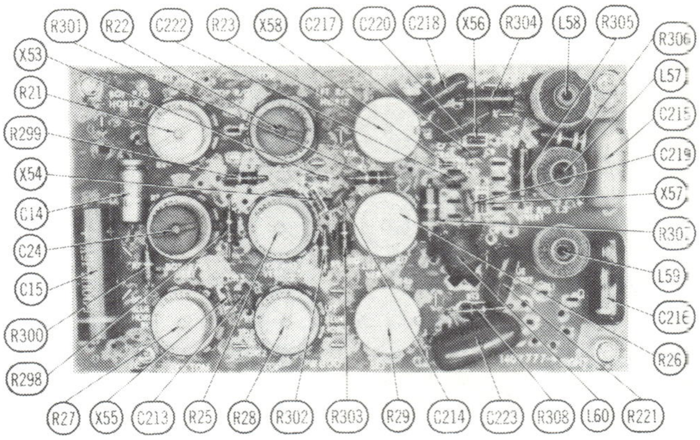

Fig. 15-14. Typical convergence panel.

CONVERGENCE ADJUSTMENTS

Step	Control	Use to Converge (or Straighten)	Remarks
1.			Perform Center Dot Convergence using convergence magnets. See Fig. A.
2.	R-G Vertical Lines, Top R25	Red and Green Vertical bars at top of screen.	Touch up both controls for best convergence from top to bottom along vertical center line (Fig. B).
3.	R-G Vertical Lines, Bottom R24	Red and Green Vertical bars at bottom of screen.	
4.	R-G Horizontal Lines, Top R22	Red and Green Horizontal bars at top of screen.	Touch up both controls for best convergence of horizontal bars along vertical center line (Fig. B).
5.	R-G Horizontal Lines, Bottom R21	Red and Green Horizontal bars at bottom of screen.	
6.	Blue Horizontal Lines, Top R28	Blue Horizontal bars at top of screen.	Touch up both controls for best convergence of horizontal bars along vertical center line (Fig. C).
7.	Blue Horizontal Lines, Bottom R27	Blue Horizontal bars at bottom of screen.	
8.			Perform Center Dot Static Convergence (Fig. A).
9.	Blue Horizontal Lines, Right L59	Blue Horizontal bars at right side of screen.	Touch up both controls for best convergence of horizontal center line (Fig. D).
10.	Blue Horizontal Lines, Left R29	Blue Horizontal bars at left side of screen.	
11.	R-G Vertical Lines, Right L57	Red and Green Vertical bars at right side of screen.	(Fig. E)
12.	R-G Horizontal Lines, Right L58	Red and Green Horizontal bars at right side of screen.	Use control to converge blue bar with red and green bars on right side of screen (Fig. E).
13.	R-G Vertical Lines, Left R26	Red and Green Vertical Bars at left side of screen.	(Fig. E)
14.	R-G Horizontal Lines, Left R23	Red and Green Horizontal bars at left side of screen.	Use control to converge blue bar with red and green bars at left side of screen (Fig. E).

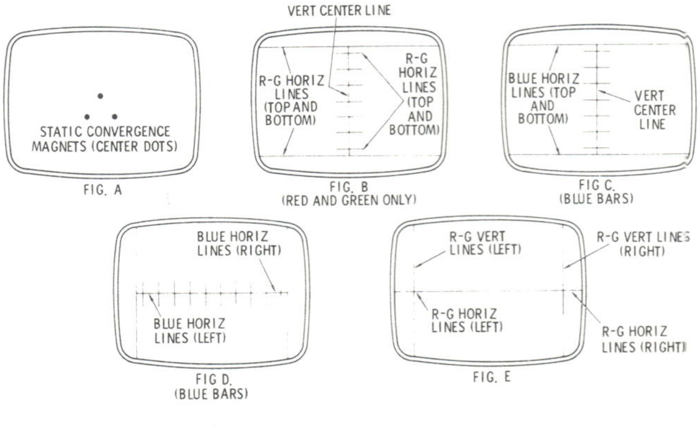

Fig. 15-15. Convergence adjustments as given in Sams PHOTOFACT®.

needing dynamic convergence. Notice that the lines are white near the center of the screen, indicating good static convergence, but they diverge into colored lines near the edges of the screen. This indicates poor dynamic convergence.

The dynamic-convergence procedure is normally carried out with a set of twelve controls located on the chassis or on the rear of the cabinet. Fig. 15-14 shows a typical convergence panel.

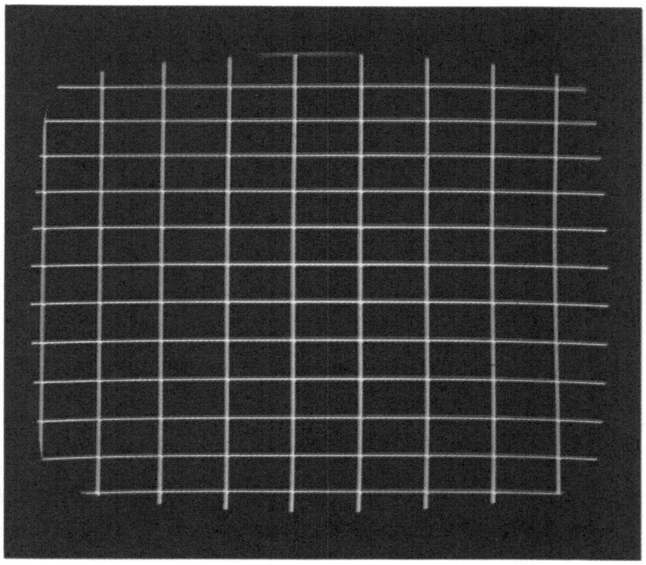

Fig. 15-16. Pattern on properly converged set.

The exact procedure to be carried out varies somewhat from one set to another. The best procedure is to purchase the Sams PHOTOFACT® folder for the set to be adjusted. An exact convergence procedure is contained in the PHOTOFACT® folder. Fig. 15-15 is a reproduction of a page from the PHOTOFACT® for the set whose convergence panel is shown in Fig. 15-13. You should study the convergence steps, identify each control, and determine the effect on the screen. When the convergence procedure has been completed, the correctly converged screen will appear as shown in Fig. 15-16.

15-8 Summary of Setup Procedures

The complete color-receiver setup procedure may be summarized as follows:

1. Set up the set for black-and-white viewing.
2. Degauss the set.
3. Adjust the purity.
4. Set the gray scale (approximately).
5. Static-converge the set.
6. Dynamic-converge the set.
7. Finally, reset the gray scale.

Do not forget to reset the gray scale as the final step in the setup procedure.

15-9 Review Questions

1. What does "setup" mean?
2. What are three categories of controls found in a tv receiver?
3. Name three black-and-white "customer" controls.
4. Explain two ways of setting the gray scale of a set.
5. Define each of the following terms:
   A. Gray scale.
   B. Static convergence.
   C. Purity.
   D. Dynamic convergence.
6. Name three "service" controls found in a black-and-white set.
7. How do you know where to put the yoke after purity setup is done?
8. Explain how static convergence is accomplished.
9. How many dynamic-convergence controls are there in most sets?
10. Tell how the purity setup procedure is carried out.
11. Name two "customer" controls found only in color sets.
12. What is a "factory" control? Name five factory controls.
13. How do you adjust the high voltage in a color set if you do not have the specific manufacturer's recommendation?
14. How do you degauss a set?
15. What is a degaussing coil?
16. Name the four controls used to perform static convergence.
17. What part of the screen do you watch while adjusting the purity rings?
18. What is a "service" control?
19. When is the service switch used?
20. Where can you find instructions on how to dynamically converge a particular set?

CHAPTER 16

Troubleshooting Color Problems

If a black-and-white receiver is thought of as a complete system, most troubles can be divided into three main categories:

1. Sound
2. Video
3. Raster

The remaining troubles that might be encountered are combinations of two or all three of these trouble categories.

On this basis, a black-and-white set can be represented in block form as shown in Fig. 16-1. In this diagram, a failure in the sound, video, or sweep blocks will show one of the symptom categories listed above. On the other hand, failures in the rf/if detector block will show combination symptoms.

16-1 Identifying Color Troubles

A color set can be looked at in the same way. Fig. 16-2 shows a block diagram of a color receiver. Notice that the color block is independent of the sound, video, and sweep blocks. For this reason, there are *four* basic symptom categories for a color set:

1. Sound
2. Video
3. Raster
4. Color.

If the sound, video, and raster are normal but the color is not, then the trouble must lie in the color block of Fig. 16-2. For this reason, the first "color" check should be to look at a black-and-white picture and listen to the sound. Any failure in black-and-white performance (or sound) is dealt with in the same way as in a black-and-white set.

A look at Fig. 16-3 reveals that a color set contains *all* of the functional blocks of a black-and-

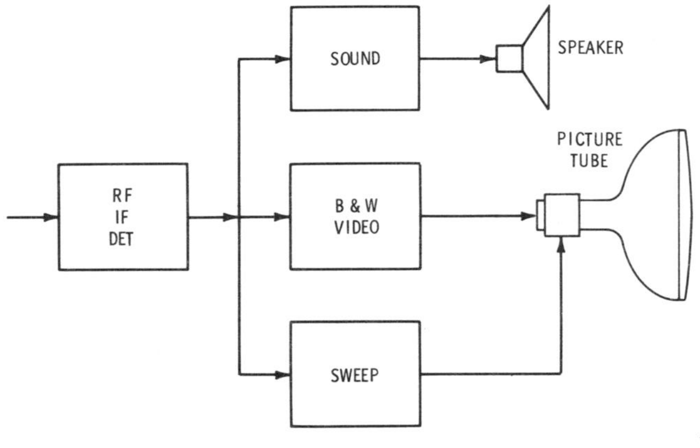

Fig. 16-1. A black-and-white receiver system.

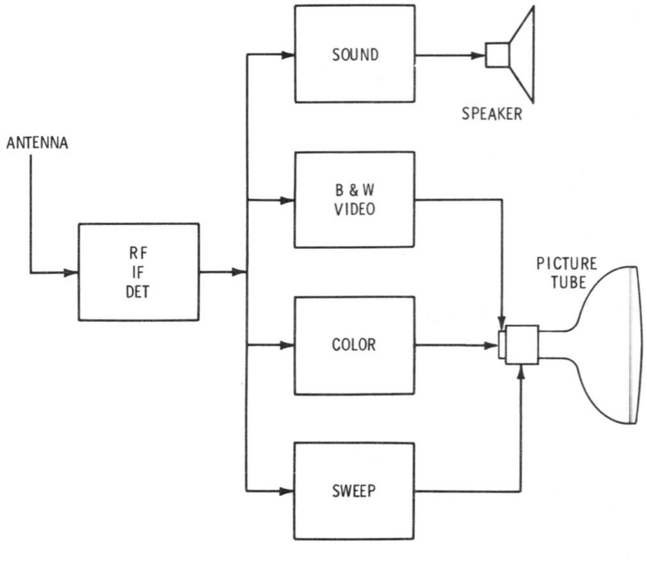

Fig. 16-2. A color receiver system.

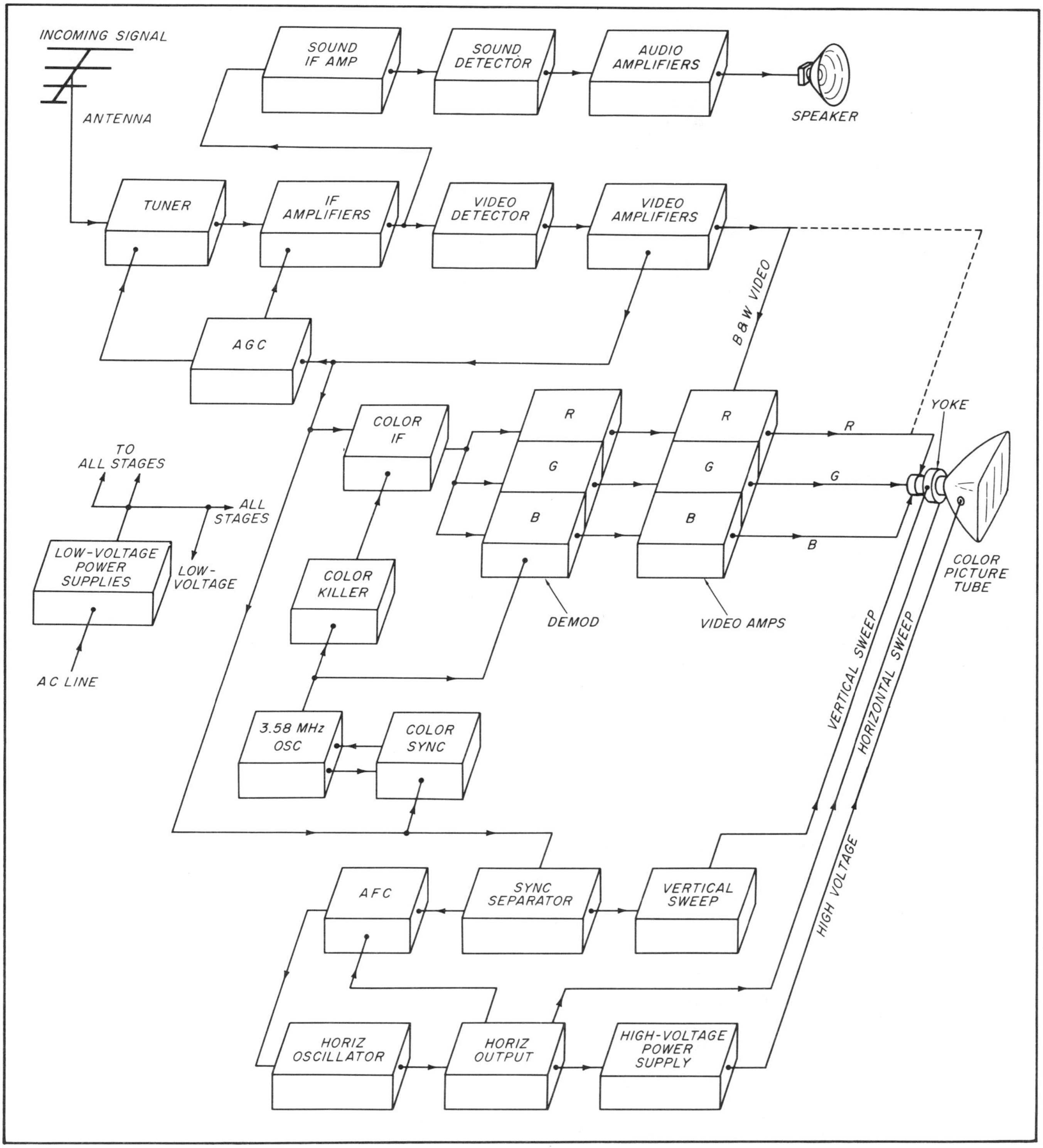

Fig. 16-3. Complete block diagram of a color receiver.

white set. It also contains six blocks of color circuitry that black-and-white sets do not have.

If the black-and-white performance (and sound) is normal but the color is not working properly, the fault must be in the color section. Take a closer look at the color section. Fig. 16-4 shows only the color blocks. From this block diagram you can get considerable insight into color problems. Go through it block by block and think about probable symptoms.

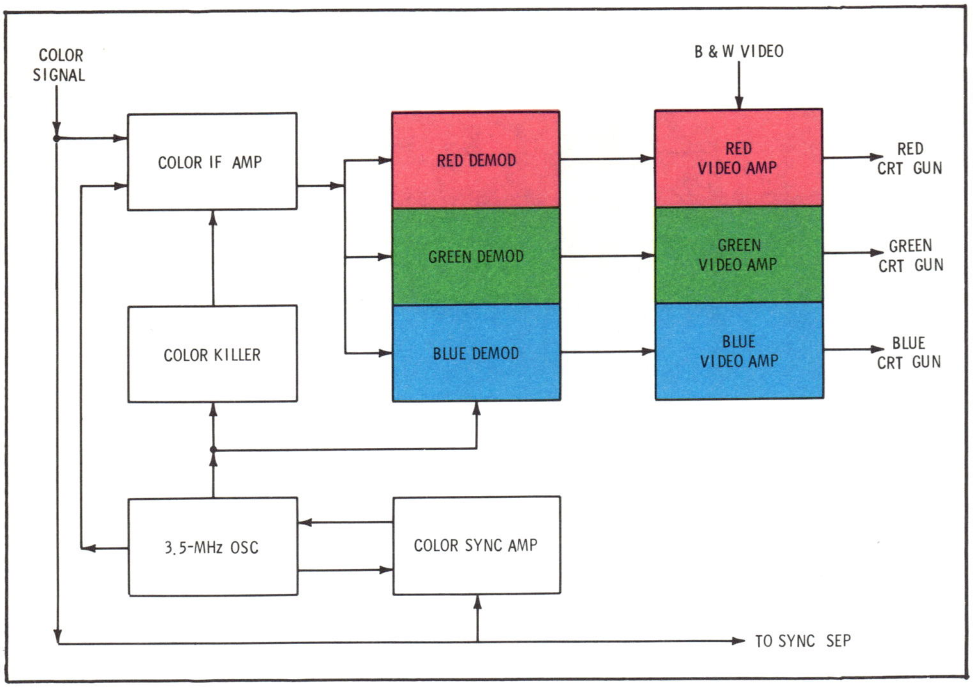

Fig. 16-4. A tv color section.

16-2 Loss of One Color

If one of the color video amplifiers (red, green, or blue) fails, then the corresponding color will be weak. For instance, if the blue video amplifier fails, the picture will be all reds, greens, and mixtures of red and green. This trouble can have an effect on the black-and-white performance, too. If the blue video amplifier is dead *and the video matrixing is done in the video circuits,* then the screen will be black and yellow instead of black and white. This symptom can also be caused by a defective blue gun in the picture tube.

There are three lost-color possibilities. One is associated with each color video amplifier. You should be able to figure out what the picture would look like if the red or green video amplifier fails.

If matrixing is done at the picture tube, then a failure of the blue amplifier will give loss of one color, but the black-and-white picture will be normal.

Failure of a color gun in the picture tube always gives loss of one color in both the color and black-and-white picture. Fig. 16-5 shows a keyed rainbow pattern and the results of the loss of red, green, or blue.

Troubleshooting the color video amplifiers is very much the same as for black-and-white video amplifiers. For example, consider the circuit shown in Fig. 16-6.

Suppose that the symptom shown in Fig. 16-5C appears and the black-and-white picture is tinted magenta. This would be diagnosed as loss of green color. First the picture tube should be checked. If it is all right, use an oscilloscope to check the green video driver and output stages. Fig. 16-7 shows the waveforms that *should* be seen with a keyed-rainbow pattern.

In this case, suppose that waveform W12 at the collector of the green driver is normal, but waveform W13 at the collector of the green output stage is missing. The conclusion would be that the problem was in the green output stage. Checking the circuit voltages with a voltohmmeter would lead to the defective component.

As you might suspect from the block diagram given in Fig. 16-4, a failure in one of the color demodulators can cause loss of one color.

16-3 One Wrong Color

If two of the colors seem normal and the third is present but has the wrong color, the demodulator should be suspected. For instance, if L1L in Fig. 16-6 were defective, the phase of the 3.58-MHz reference signal would be shifted. This would cause a change in blue content of the picture. Again a waveform check will reveal the problem. In this case, waveform W8 in Fig. 16-7 would be present, but it would not have the correct shape. To a certain extent, these troubles may be cleared

(A) Normal keyed rainbow pattern.

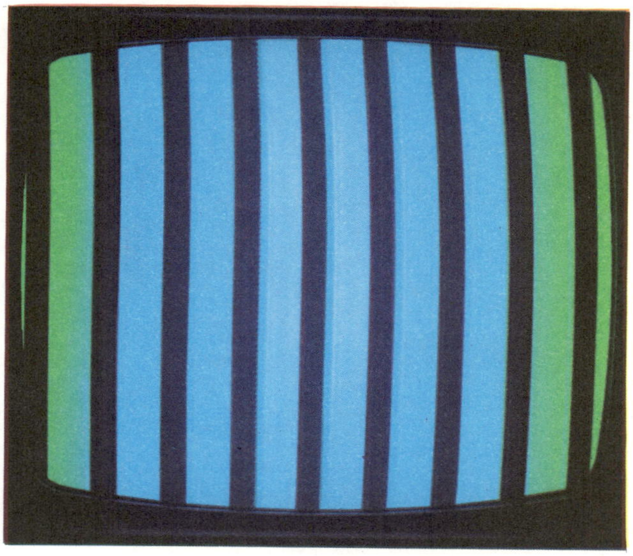

(B) No red.

(C) No green.

(D) No blue.

Fig. 16-5. Loss of one color.

by readjustment of the hue control; however, if much change is required, a definite trouble exists.

16-4 No Color

So far, problems that affect only one color have been discussed. Now look at things that can completely kill the color. Suppose you are working with the color pattern shown in Fig. 16-8A and the screen looks like Fig. 16-8B. You would call this *no color*. This symptom can be caused by any of the remaining four stages in the color section. Referring to Fig. 16-9 (which is the same as Fig. 16-4), you can see that if the color if amplifier is dead, no color signal will get through. Similarly, if the color killer holds the if stages cut off, the color will be dead. If the oscillator is dead, the color killer will cut off the if stages. Also if the oscillator is dead, there will be no 3.58-MHz reference signal to the demodulators. Finally, the oscillator will not operate without a burst signal from the color-sync amplifier. The overall result is that all four stages must work to give a color picture.

Let's examine a typical circuit and discuss troubleshooting a no-color problem. Fig. 16-10 shows a color-sync amplifier and 3.58-MHz oscillator circuit. The normal troubleshooting procedure for a no-color symptom is to start at the input of the

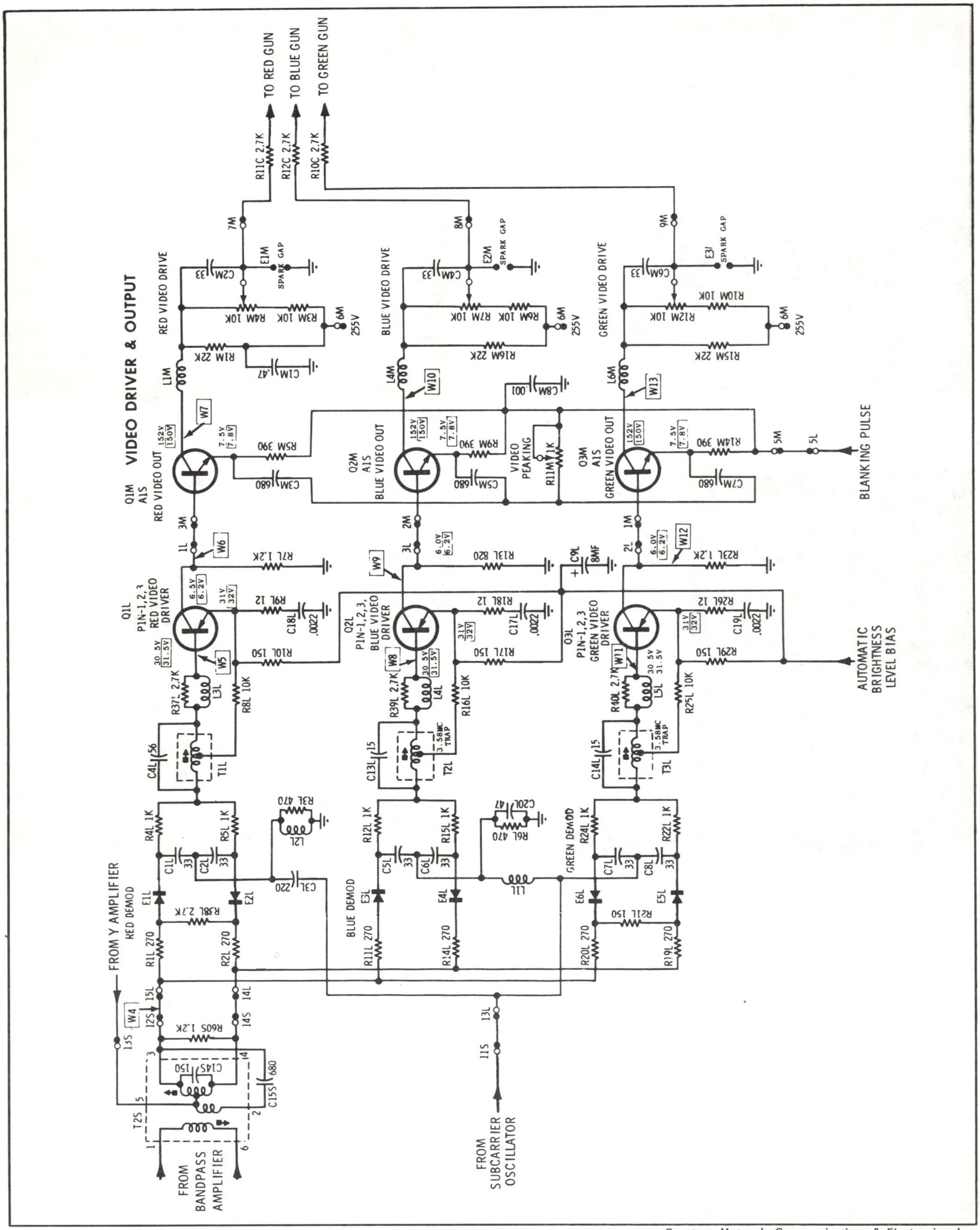

Courtesy Motorola Communications & Electronics, Inc.

Fig. 16-6. Color video amplifiers.

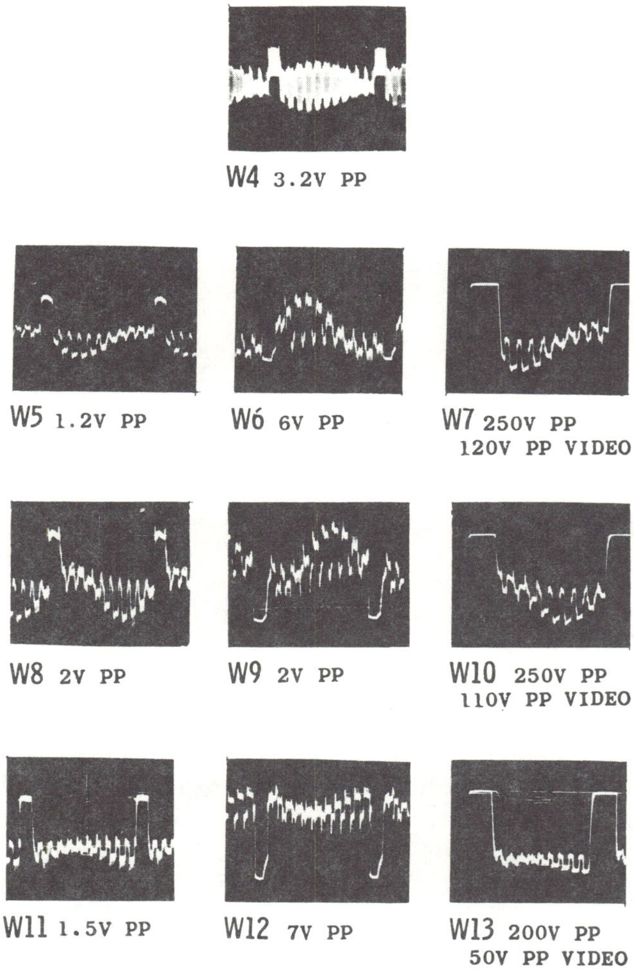

Courtesy Motorola Communications & Electronics, Inc.

Fig. 16-7. Waveforms for Fig. 16-6.

color-sync amplifier and view the waveforms at each stage. In making these tests, a color generator should be connected to the set. The waveforms shown in the illustration were taken with a keyed rainbow pattern.

When one of the waveforms is found to be missing, the associated circuitry is checked with a volt-ohmmeter. In this way, the defective component can be identified.

The color if and color-killer circuits shown in Fig. 16-11 are handled in the same way. Notice that waveforms are not shown for the color-killer circuit. This stage operates principally on dc levels, so a voltage check alone is enough.

It will probably be no surprise that we can also have color that is weak. Weak color will give a symptom somewhere between those shown in Figs. 16-8A and 16-8B. While it is possible that any one of the four stages (color sync, color oscillator, color killer, or color if) could cause weak color, the basic problem will most likely be low amplification in the color if (bandpass amplifier) section. Check the signal levels as before, looking for the cause of the reduced amplification.

(A) Normal NTSC color pattern.

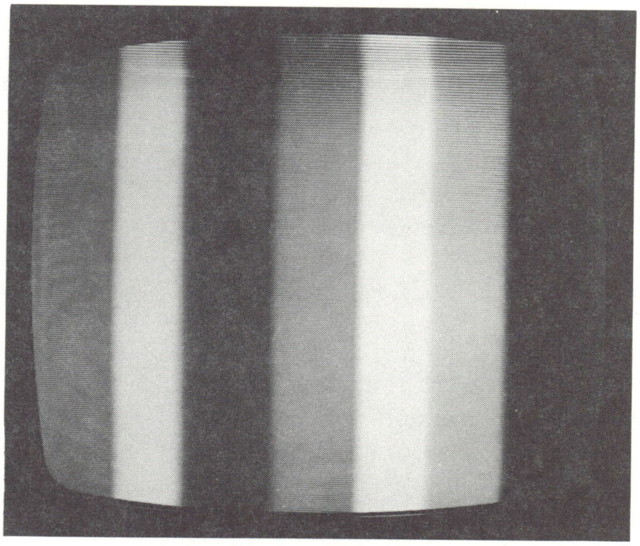

(B) NTSC pattern with no color.

Fig. 16-8. Appearance of no-color condition.

16-5 Color Too Intense

The opposite extreme of no color is color that is too intense. Fig. 16-12 shows an example of this symptom. If the color-intensity control cannot correct this problem, then you must conclude that the color if amplifier has too much gain. This can be checked by viewing the input and output of the color if amplifier with an oscilloscope. The problem itself may lie in the color if amplifier, in the

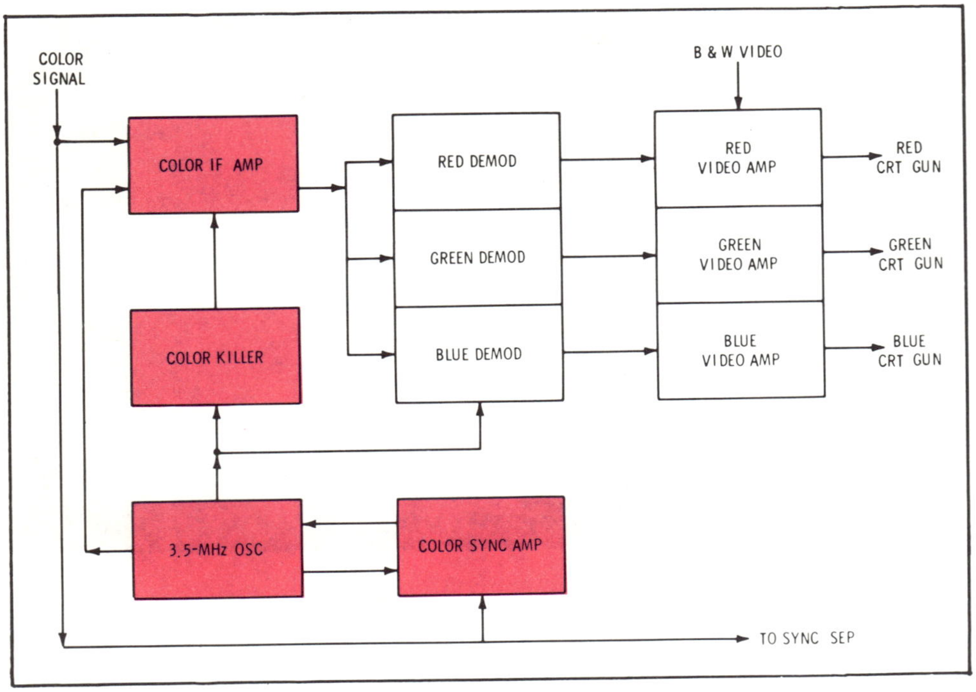

Fig. 16-9. A tv color section.

color killer, or in the automatic chroma-control stage (acc). These last two stages have some measure of control over the color if gain. They can usually be checked easily with a voltohmmeter.

16-6 All Colors Wrong

It is possible for a set to present all of its colors incorrectly. Compare Fig. 16-13A with Fig. 16-13B, and you will notice that all of the bars in Fig. 16-13B are the wrong color. The *sequence* of the colors is, however, correct. You can produce this symptom simply by misadjusting the *color* (or hue) control. If you refer to Fig. 16-10, you will see that the hue control (R59S) actually shifts the phase of the color-oscillator output signal.

Therefore, it can be seen that any failure (such as C38S changing value) that causes a phase shift in the oscillator signal will cause these symptoms. The fault may easily be severe enough that the hue control cannot correct it.

Also, any failure in the color if amplifier that causes phase shift in the color signal can cause this symptom. Troubleshooting this kind of if failure is outside the scope of this coverage, since it requires a knowledge of color if alignment.

16-7 Loss Of Color Sync

We are all familiar with an out-of-sync picture in a black-and-white set. You might expect loss of color sync to look something like loss of horizontal sync. Actually the two are quite different. Fig. 16-14 shows both types of loss of sync. Notice that the color bars in Fig. 16-14A are in sync horizontally, but each is multicolored. This is the classic symptom of *loss of color sync* with a keyed rainbow pattern.

As you know, the horizontal-oscillator frequency must be correct if a black-and-white picture is to be in sync. In the same way, the 3.58-MHz color oscillator must have the correct frequency if the color picture is to be in color sync.

Consequently, the color oscillator should be suspected when color sync is lost. Troubleshoot the stage as you would a horizontal oscillator; that is, view the waveform and check components and voltages in the oscillator circuit. In the oscillator circuit shown in Fig. 16-10, you should suspect the transistors, the crystal, and the tuned circuits.

In addition to operating at the correct frequency, the color oscillator must be phase-locked with the color burst from the station. A failure in the color-sync amplifier or gating circuit can cause loss of phase lock. The result can be loss of color sync. Again, the best troubleshooting procedure is to check waveforms and circuit voltages.

16-8 Summary of Color Symptoms

So far, six basic color symptoms have been discussed. These symptoms and the stages that nor-

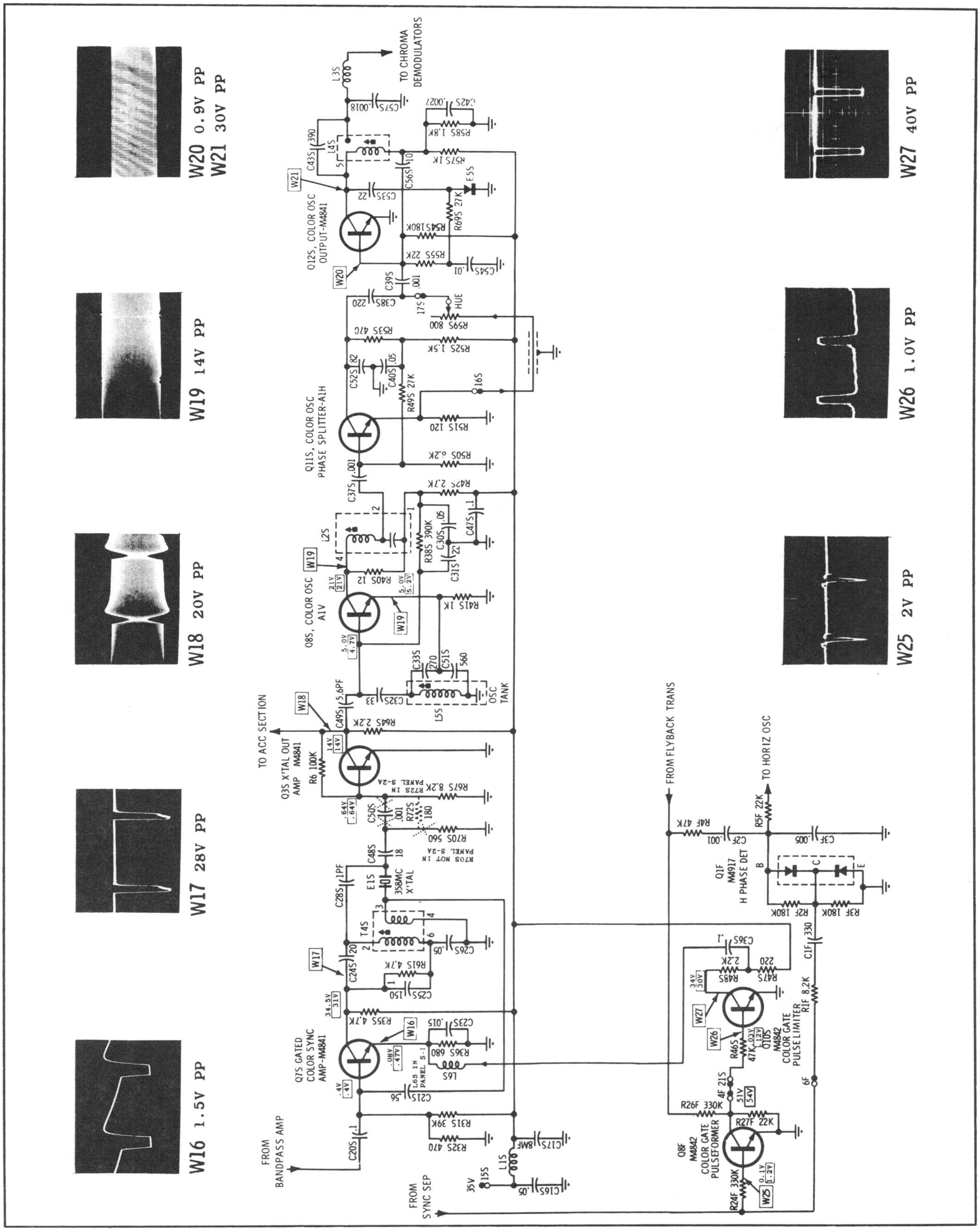

Fig. 16-10. Color-sync amplifier and 3.58-MHz oscillator.

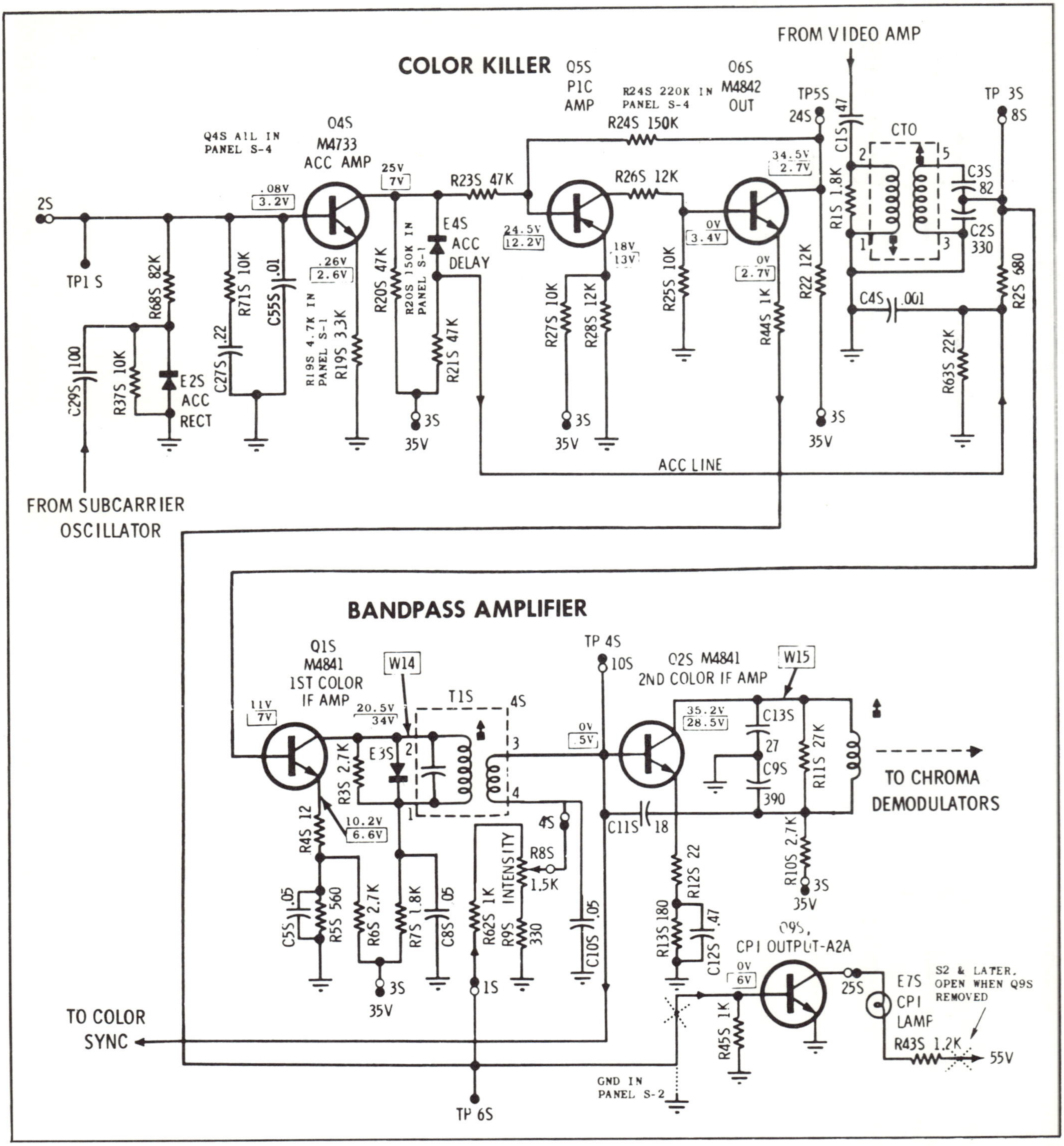

Fig. 16-11. Color killer and color if amplifier.

mally cause them are summarized in Table 16-1. Before checking color symptoms be sure that black-and-white performance is normal.

Troubleshooting the color circuitry is no more difficult than troubleshooting black-and-white circuitry. The most effective procedure is to view waveforms to find the stage causing the trouble. Then measure circuit voltages to indicate suspected components. Finally, check components to identify the one that has failed.

Table 16-1. Basic Color Symptoms and the Stages Involved

Symptom	Suspected Stages
Loss of One Color (See Fig. 16-5.)	Picture Tube Color Video Amp Color Demodulator (Check drive and screen controls first.)
One Wrong Color	Color Demodulator
No Color (See Fig. 16-8B.)	Color IF Amp Color Killer 3.58-MHz Osc Color Sync Amp (Check color killer control first.)
Color too Intense (See Fig. 16-12.)	Color IF Amp Automatic Chroma Control Color Killer (Check color intensity control first.)
All Colors Wrong. (See Fig. 16-13B.)	3.58-MHz Osc Color IF Amp (Check hue control first.)
Loss of Color Sync (See Fig. 16-14A.)	3.58-MHz Osc Color Sync Amp Color Sync Gate

16-9 Review Questions

Use the circuit diagrams, photos, and summary of this chapter to answer the following questions.

1. What is the name of the symptom in Fig. 16-15?

Fig. 16-12. Color too intense.

(*A*) *Normal keyed-rainbow pattern.*

(*B*) *All colors wrong.*

Fig. 16-13. Appearance of screen with all colors wrong.

2. Which set control might correct this problem?
3. Referring to the circuit in Fig. 16-10, which stages would you suspect of causing the symptoms shown in Fig. 16-15? Explain why.
4. Which stages in Fig. 16-6 do you think would cause the symptom shown in Fig. 16-16? Explain your answer.
5. Which color is affected in Fig. 16-16? How can you tell?
6. In Fig. 16-7, waveform W6 is normal and W7 is missing. Which symptom in Figs.

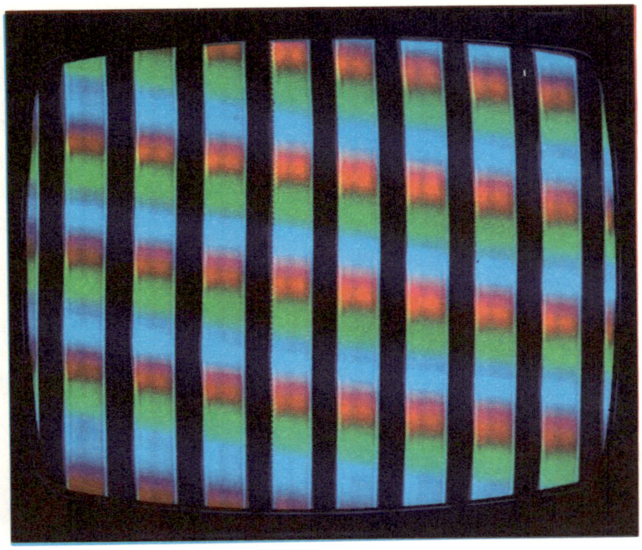

(*A*) *Loss of color sync.*

(*B*) *Loss of horizontal sync.*

Fig. 16-14. Appearance of loss of sync.

16-15 through 16-22 would you expect? Why?
7. How would you proceed to find the defective component in Question 6?
8. What do you call the symptom shown in Fig. 16-21? In what section of the set would you expect the fault to lie?
9. If you observe the symptoms shown in Fig. 16-17, which stages of the set would you suspect?
10. The tuning slug in L5S of Fig. 16-10 is missing. Do you think this could cause the symptoms in Question 9? Why?
11. Which stages would you check if you had

Fig. 16-15. Color symptom A.

Fig. 16-16. Color symptom B.

Fig. 16-17. Color symptom C.

Fig. 16-18. Color symptom D.

Fig. 16-19. Color symptom E.

Fig. 16-20. Color symptom F.

Fig. 16-21. Color symptom G.

Fig. 16-22. Color symptom H.

a set with the symptoms shown in Fig. 16-18?

12. Waveform W15 in Fig. 16-11 is eight volts peak to peak. Normally, it should only be four volts peak to peak. Which of the symptoms in Figs. 6-15 through 16-22 would you expect?
13. Inductor L1L in Fig. 16-6 is open-circuited. Which of the symptoms in Figs. 16-15 through 16-22 do you think this would cause? Why?
14. The symptom shown in Fig. 16-19 is observed on a set when an NTSC color pattern is used. How would you troubleshoot this problem?
15. Is there a control that could cause the symp-

toms shown in Fig. 16-19? If so, what is it?

16. Name each of the symptoms in Figs. 16-15 through 16-22.
17. Are all of the symptoms shown in Figs. 16-15 through 16-22 caused by problems in the color section? Explain your answer.
18. Which control on a set could cause the symptom shown in Fig. 16-18?
19. Could the symptom shown in Fig. 16-22 be caused by improper gray-scale adjustment? Explain your answer.
20. Which of the symptoms in Figs. 16-15 through 16-22 could be caused by a defective picture tube? Tell why you think a picture tube could cause each of the symptoms you list.

CHAPTER 17

Special Receiver Features

In previous chapters, you have already encountered two special receiver features. Automatic gain control was discussed in Chapter 8. Automatic chroma control was mentioned in Chapter 16. Since it was considered only briefly, it is perhaps worthwhile to reconsider it now.

17-1 Automatic Gain and Chroma Control

As you will recall, agc and dagc are used to maintain the same picture quality for both strong and weak stations. Automatic chroma control (acc) does the same thing for the color signal.

The color burst is usually taken off at the output of the burst amplifier (or color sync amplifier). It is detected by a diode circuit called a color detector (or acc detector). Sometimes the detected acc signal is amplified in a stage called the acc amplifier. The acc signal is then used to control the color-killer stage which, in turn, controls the gain of the color if amplifiers. Fig. 17-1 shows a typical acc system of this type.

In some other sets, the automatic chroma control signal is applied directly to the color if stages instead of going through the color killer. Fig. 17-2 shows a system of this type.

In some other sets, both of these types of acc are used, as shown in Fig. 17-3. When this combined acc is used, the signal to the color if stage is usually "delayed" in the same manner as dagc to the tuner.

Look at an actual example of acc circuitry. Fig. 17-4 shows part of the color section of a solid-state set. Notice that the acc signal is taken off at the collector of Q3S. From the acc amplifier, the signal is applied directly to the base of Q5S, the color-killer amplifier. It is also fed through the acc delay (E4S) to the first color if amplifier (Q1S).

If the acc system fails, you will get either of two symptoms. Both of these symptoms are shown

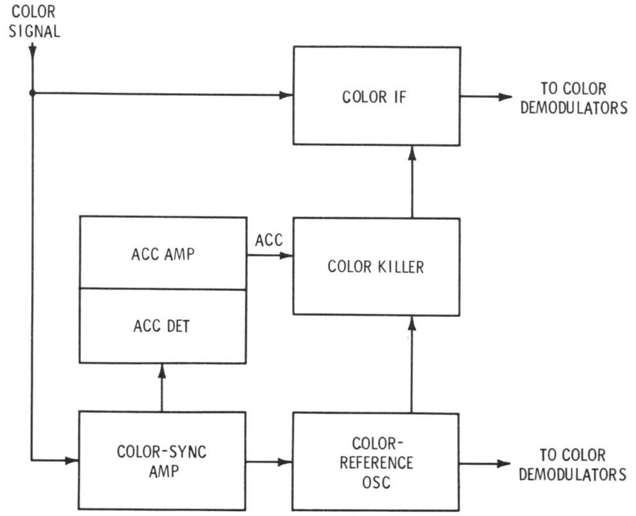

Fig. 17-1. Acc applied to the color killer.

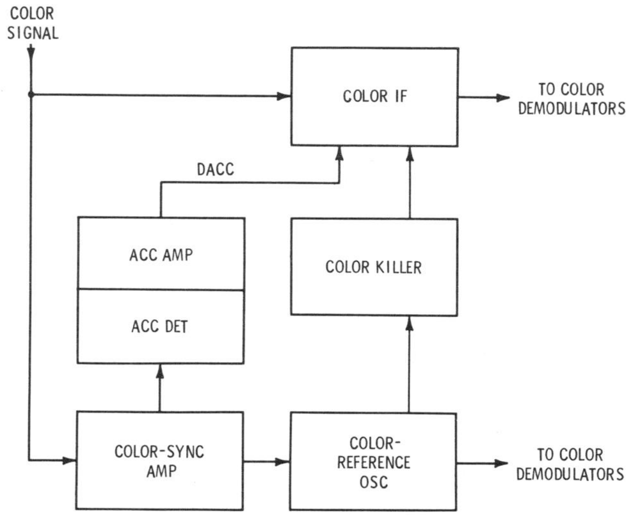

Fig. 17-2. Acc applied to color if stages.

in Fig. 17-5. If the acc fails in such a way as to completely cut off the color killer or color if amplifier, the result will be no color at all. If the acc fails in a way that holds the color section fully on, then you will have excessive color in the picture. If there is no color signal present and the color if is turned on, you will see colored snow (called *confetti*) on the screen.

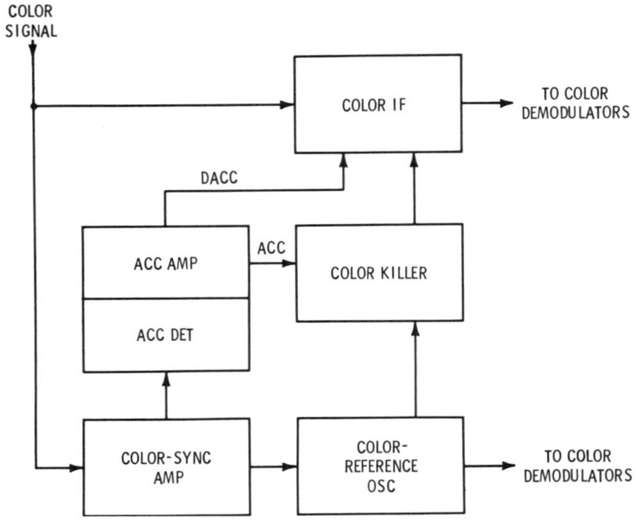

Fig. 17-3. Acc applied to the color killer and if stages.

In addition to acc, many sets also have a one-button color-preset feature. This feature allows the brightness, contrast, hue, and tint controls to be preset for a satisfactory picture. The picture can then be adjusted to some extent with color level and tint controls on the front of the set.

Recently, many broadcast stations have begun transmitting a *vertical interval reference* (*VIR*) signal. This signal can be used in the receiver for automatic adjustment of the color and tint of the crt picture to match that broadcast by the station.

17-2 Automatic Brightness Limiting

Modern color picture tubes are capable of producing bright pictures. In fact, it is possible in some cases to damage the picture tube or other set components if the picture-tube beam current becomes excessive. Consequently, many sets have an *automatic brightness limiter* (abl) circuit to prevent excessive beam currents. Fig. 17-6 shows a typical abl system.

The person watching the set adjusts the brightness to the desired level manually as usual. As the video level varies, so do the beam currents; and as the beam currents vary, the focus voltage on the picture tube also changes. This change in focus voltage is used to limit the beam currents and thus the brightness.

Focus-voltage changes are fed to the base of Q5L, the abl driver. Any increase in focus voltage tends to increase the current through this stage. This increase in abl-driver current tends to increase the current through the brightness-control transistor (Q4L), which in turn reduces the three video-driver currents through Q1L, Q2L, and Q3L. These three transistors control the actual crt beam currents, so if the beam current starts to increase, the abl system holds it down. The 8-μF capacitor in the emitter circuit of Q4L slows down the action of the abl system so that it does not respond to video changes in brightness but does respond to changes in the average brightness level.

In this system, any change in the value of the 35-volt collector voltage would change the brightness. To prevent this, Q7E, the brightness stabilizer, is used. This stage senses changes in the 35-volt supply and adjusts the emitter voltage of Q5L to keep the average brightness constant.

17-3 Automatic Fine Tuning

In the early days of color television, you had to get up every few minutes and retune the set to get good color reception. Most modern sets now do this automatically. Fig. 17-7 shows a typical aft (automatic fine tuning) system. In this system, the aft circuit monitors the output of the third video if stage. If the frequency of the if signal is 45.75 MHz, the aft does nothing. If the if signal is not 45.75 MHz, then the aft circuit produces an output that changes the frequency of the oscillator in the tuner. In this way, the aft circuit corrects the oscillator frequency so that the intermediate frequency (if) is always 45.75 MHz.

Fig. 17-8 shows an aft circuit employing an integrated circuit (IC). The if signal is fed into the aft buffer amplifier through the tuned circuit consisting of C1301 and L1301. The amplified output of this stage is fed to the discriminator transformer, L1303. The transformer feeds the signal into the second half of the IC, which contains the discriminator stage and an amplifier. The output of this section is fed to both the vhf and uhf tuners, where it is used to correct the oscillator frequencies.

The defeat switch at the output of the aft circuit is used to kill the aft action when the set is being manually fine-tuned. It also works momen-

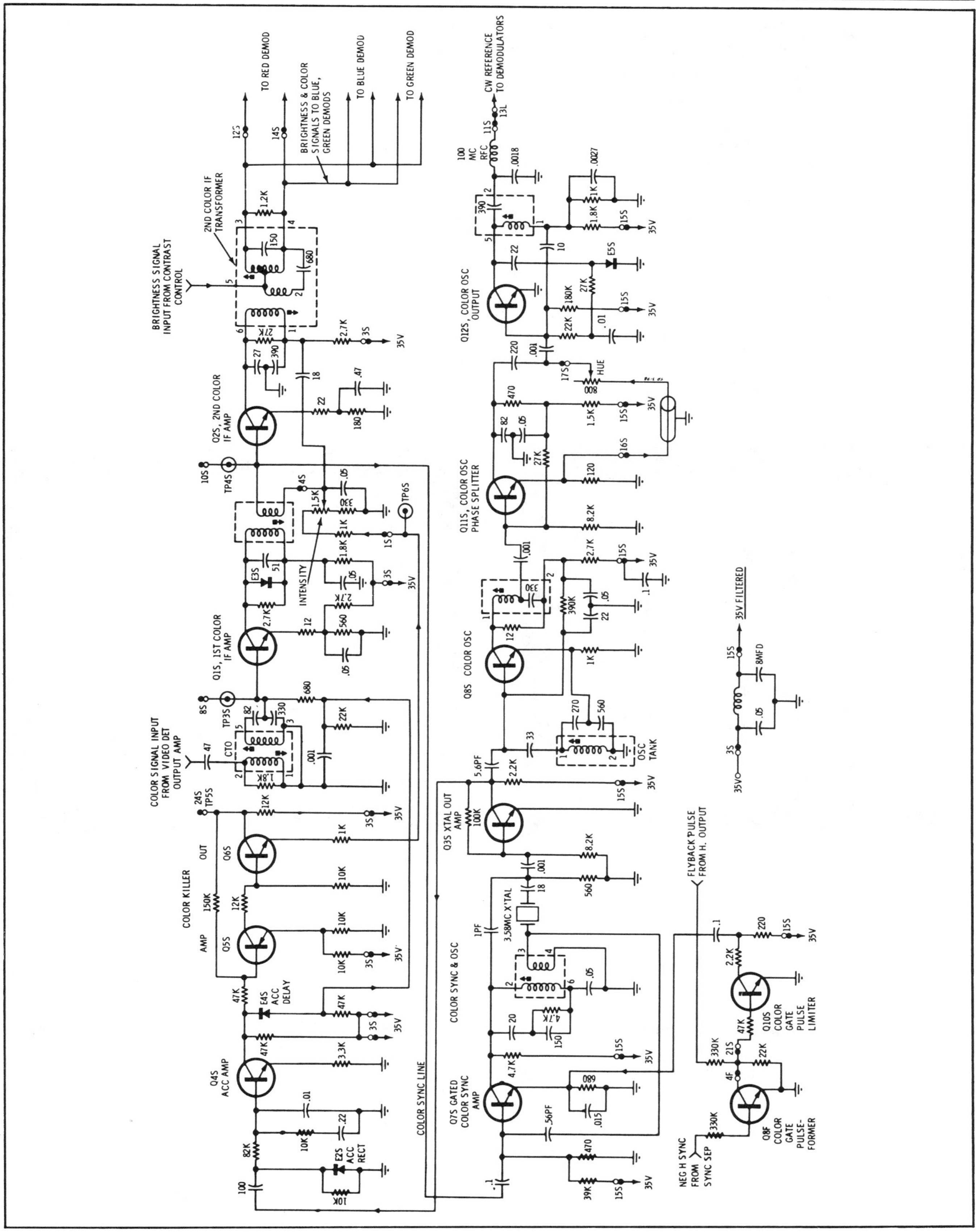

Courtesy Motorola Communications & Electronics, Inc.

Fig. 17-4. A typical solid-state system.

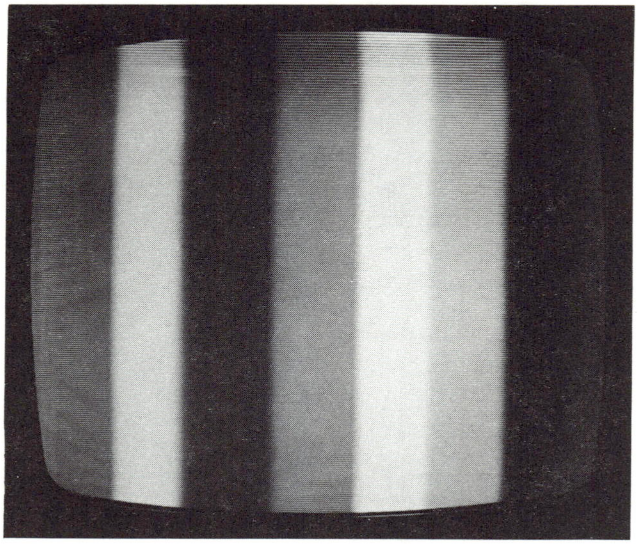

(*A*) *No color.*

(*B*) *Excessive color.*

(*C*) *Confetti.*

Fig. 17-5. Acc failure symptoms.

tarily when the channel selector is being switched from one channel to another.

17-4 Fine-Tuning Indicators

Some sets have a circuit that is used to indicate when the set is properly fine-tuned. These circuits usually light up an indicating lamp or deflect a shadow meter when fine-tuning is needed. A few use a regular meter movement that indicates when the fine-tuning is peaked. Fig. 17-9 shows a typical fti (fine-tuning indicator) circuit.

As in the case of aft, this circuit monitors the output of the video if section. Diodes D1 and D2 form the frequency-detector circuit. When the intermediate frequency drifts away from 45.75 MHz, a signal voltage is produced. This signal is amplified by Q1, Q2, and Q3 and turns on the indicator lamp, E11 (or E10). When the intermediate frequency is 45.75 MHz, no signal is produced, and the lamp is out. Some sets use both an aft and an fti circuit, but others have only one or the other.

17-5 Line-Voltage Regulators

Television receivers have always been more or less sensitive to variations in line voltage. With the advent of solid-state sets, the problems have

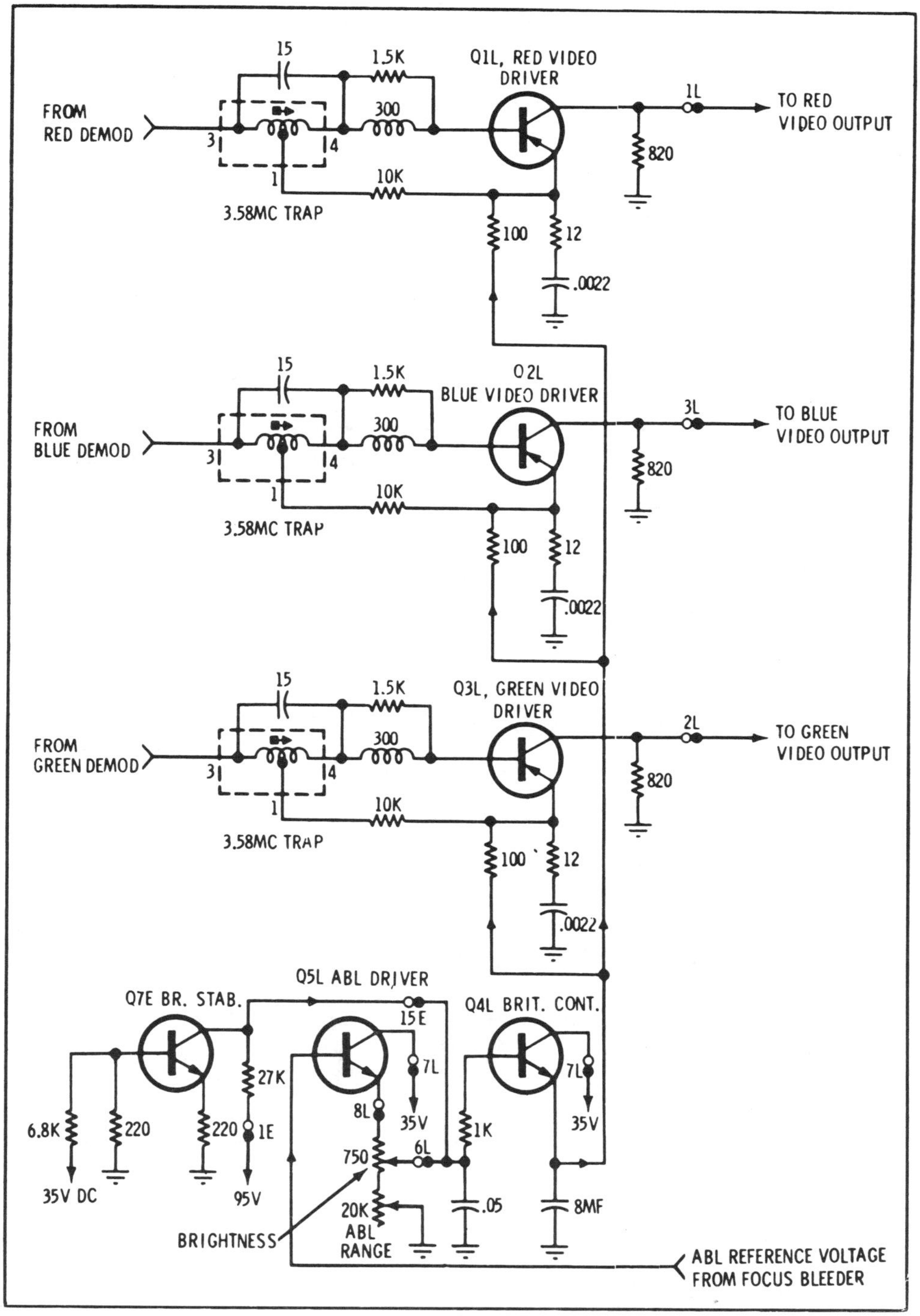

Fig. 17-6. An abl system.

Courtesy Motorola Communications & Electronics, Inc.

become somewhat more acute. As a result, some sets now have a circuit that regulates the voltage of the 120-volt ac line to the input to the set.

Fig. 17-10 shows the diagram of one of these circuits. With this circuit, the input can vary from 105 volts up to 125 volts, and the voltage applied to the set stays almost constant. Basically, the circuit is an electronic switch that is synchronized with the input line voltage. Fig. 17-11 shows a block diagram that will help in understanding the operation of the circuit.

The power transformer, T1, is in series with a 25-ohm resistor that drops the transformer voltage to 105 volts when the line voltage is 125 volts. Across the resistor is a *triac*. This device is simply an electronically controlled switch. The triac is

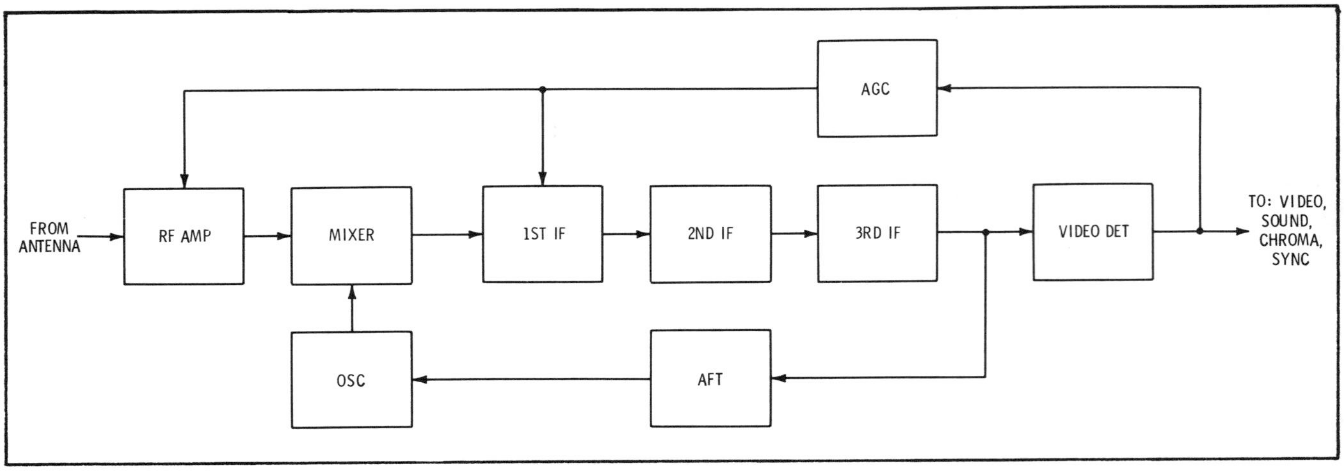

Fig. 17-7. Block diagram of an aft system.

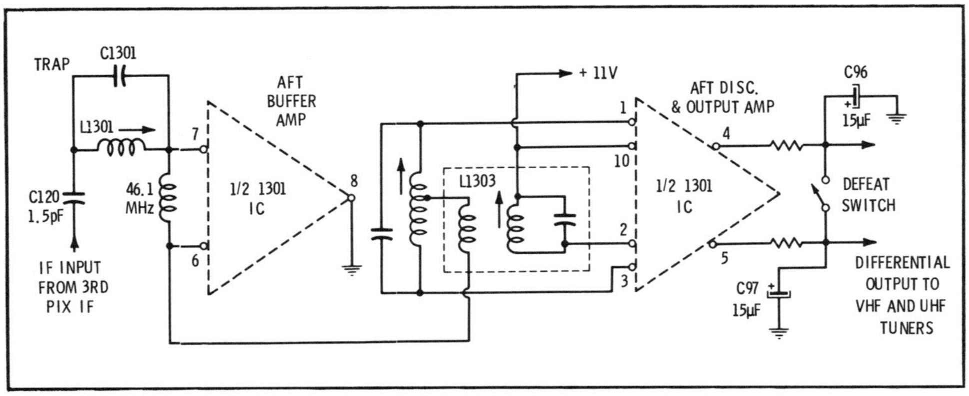

Fig. 17-8. Circuit diagram of an aft system.

opened and closed as necessary every half of the input cycle by the regulator amplifier and sync circuit.

The amplifier circuit monitors the B+ voltage of the set. If the B+ voltage is a little too high, the amplifier keeps the triac switch open. This lowers the transformer voltage to 105 volts. If the B+ is a little too low, the amplifier closes the triac switch during part of each input cycle, raising the transformer voltage above 105 volts. In this way, the regulator holds the B+ voltage in the set constant.

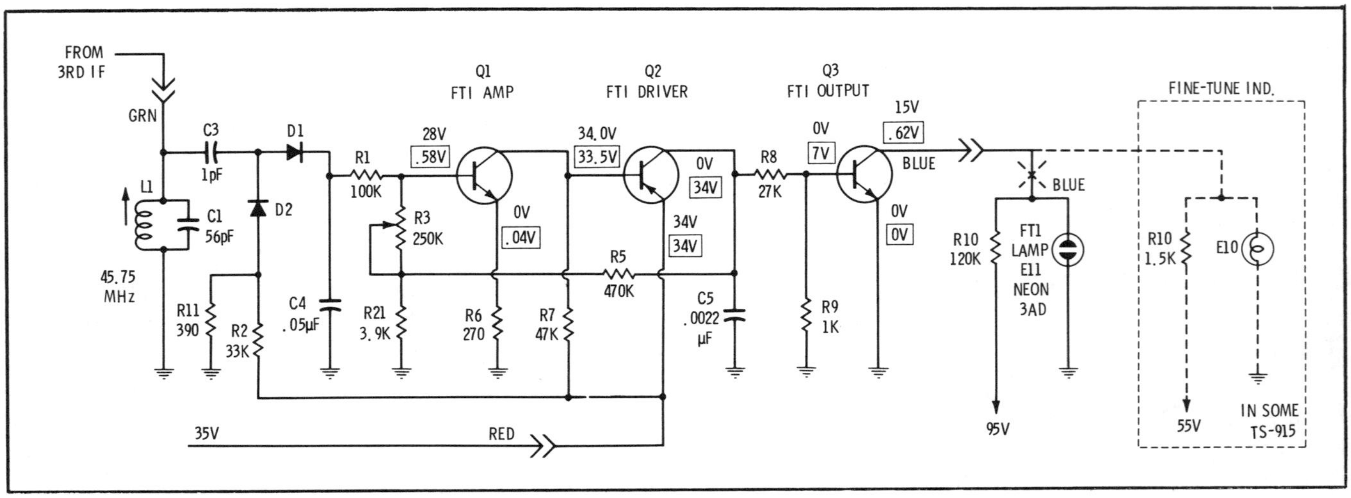

Fig. 17-9. An example of an fti circuit.

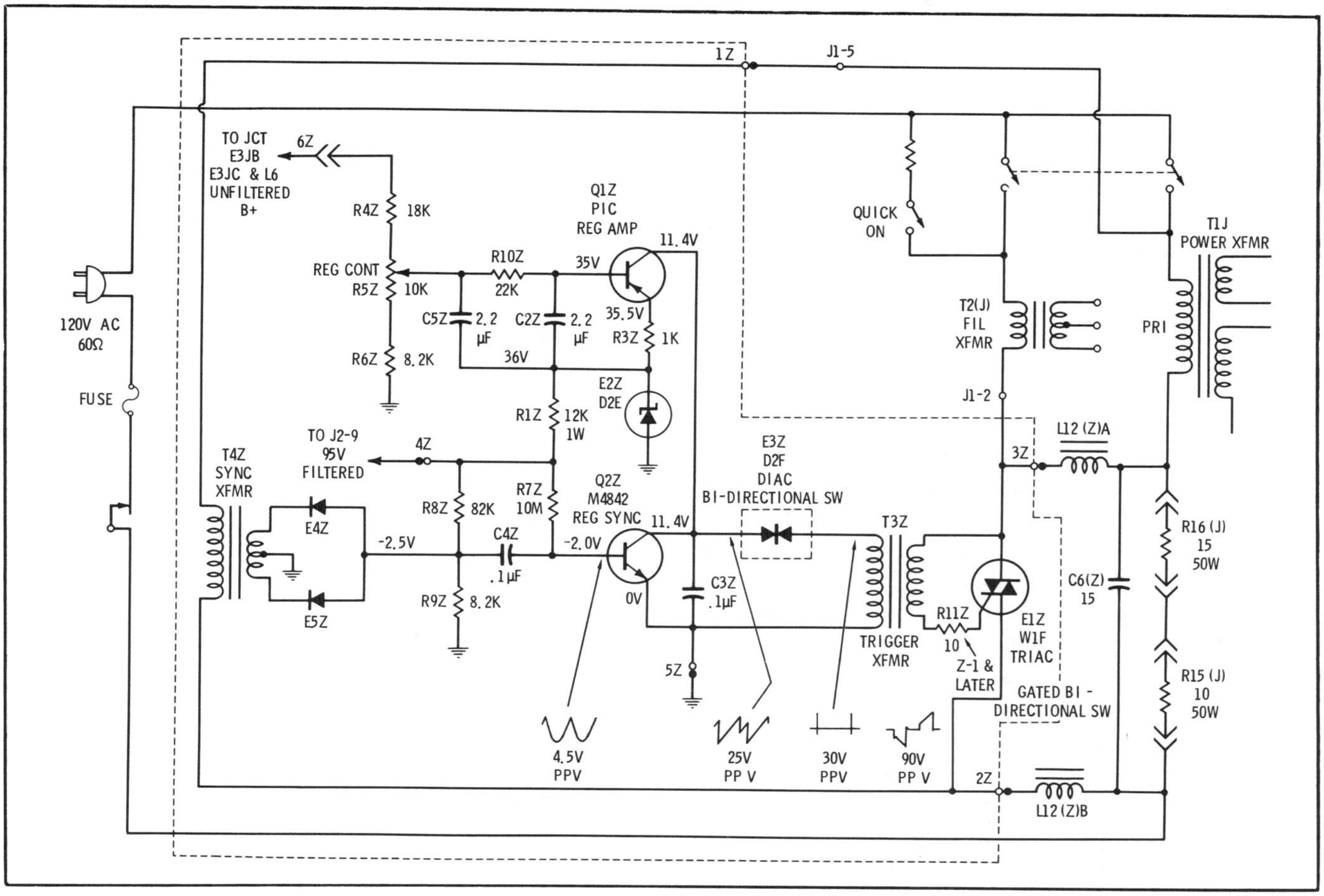

Fig. 17-10. Circuit of an ac line-voltage regulator.

There are several other types of line regulators that could be used in tv receivers, but at present this type is the one most often encountered.

17-6 Troubleshooting Special-Feature Failures

A failure in a special feature is usually not very hard to identify. However, troubleshooting them can be quite confusing if you do not have good service information. The best advice is to consult reliable service data for the particular set. Examine the circuit diagram of the suspected circuit. Then check the circuit voltages and waveforms for abnormal indications. This will normally lead you to the defective circuit and then to the component that has failed.

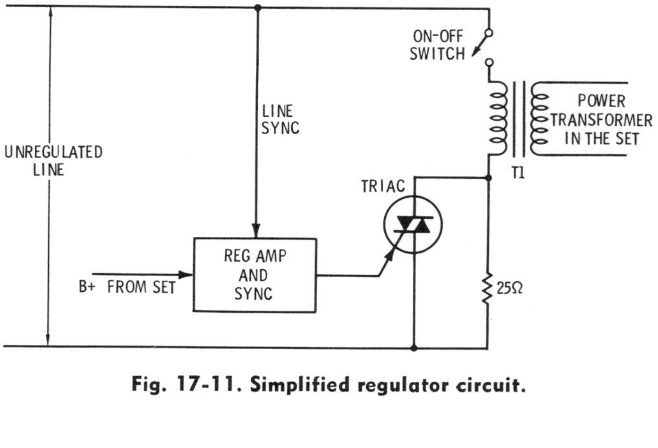

Fig. 17-11. Simplified regulator circuit.

17-7 Basic Remote-Control Operation

As more and more automatic features are added to the modern receiver, remote control becomes more and more convenient. In the old days when most receivers required readjustment every time you changed channels, a remote control was of limited value. Today, a set needs corrective adjustments only rarely, so remote controls are increasing in popularity.

Television remote controls all work on basically the same idea. Fig. 17-12 shows a typical system. You can get an idea of how the system works by going through the block diagram from left to right.

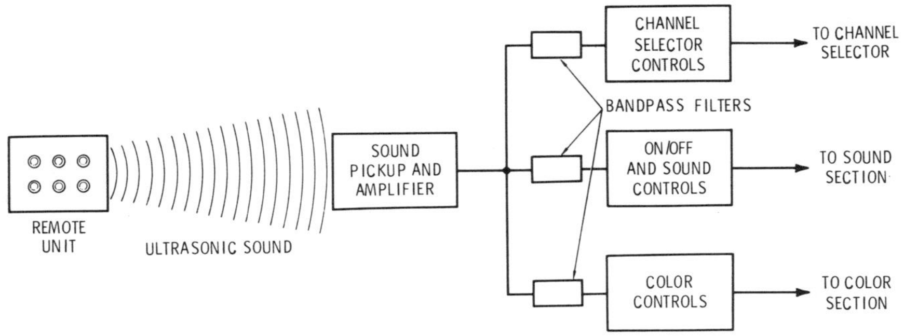

Fig. 17-12. A remote-control system.

The remote unit on the left is a transmitter that produces ultrasonic sound waves. The transmitter itself may be either mechanical or electronic. We shall look at each type in later paragraphs. The output signals are too high in frequency to be heard, but you can hear the snap of the mechanisms in the mechanical units. The transmitter may produce three or more different output frequencies, but they will all usually be near 40 kHz.

The ultrasonic output signals travel through the air as sound waves to the sound pickup. The pickup itself is in effect an ultrasonic microphone that "hears" the transmitter output. After the pickup converts the ultrasonic signal to an electronic one, the signal is amplified and fed to the bandpass filters.

Each bandpass filter is tuned to pass only one of the transmitter output frequencies, and that signal actuates one particular control function. For example, suppose we press the channel-select button on the remote-control transmitter. Further suppose that the transmitter then produces a 41.25-kHz output. This signal is picked up by the receiver and amplified before being applied to the bandpass filters. Only the channel-selector filter is tuned to 41.25 kHz. As a result, only the channel-selector controls are actuated. The other remote operations would be tuned to different frequencies.

17-8 Mechanical Remote Transmitters

One of the popular types of remote-control transmitters in use today employs mechanical tuning elements. Fig. 17-13 shows one of these transmitters. When one of the remote-control buttons is pressed, a small spring-loaded striker is pulled back and released much like the hammer of a toy pistol. After being released, the striker snaps forward and hits the end of a tuned rod. When struck, the rod "rings" much like a bell at a frequency determined by its length. The ringing signal from these rods carries 40 or 50 feet (or even more, depending on obstructions, etc.) with enough strength to actuate the receiver.

The main advantages of the mechanical type of transmitter are that it is not likely to change tuning much, and it does not require electrical power. You might say it is "finger-powered."

17-9 Electronic Remote Transmitters

Electronic remote-control transmitters are transistor oscillators capable of operating at several frequencies in the ultrasonic range. Fig. 17-14 shows the circuit diagram of a four-function transmitter. Notice that the 1.5-volt battery is connected into the circuit only when one of the function buttons is pressed. Connecting the circuit in this way keeps the oscillator off, and there is no

(A) Front view. *(B) Inside view.*

Courtesy RCA

Fig. 17-13. A mechanical transmitter.

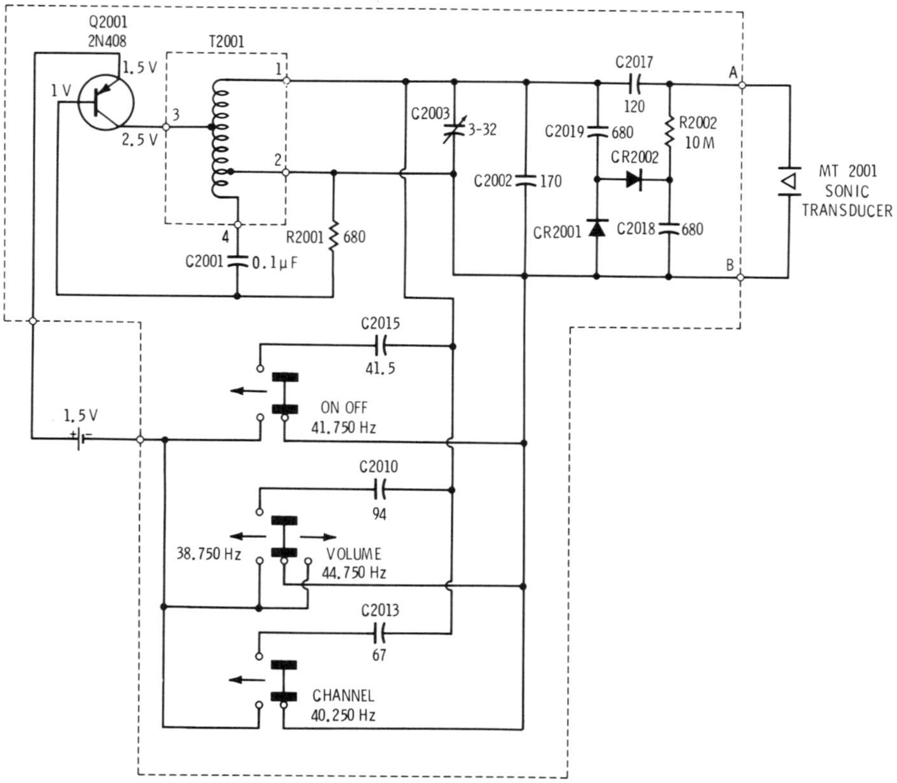

Fig. 17-14. A four-function transmitter.

battery drain until an output is desired. This results in long battery life.

Feedback in the oscillator circuit is through the oscillator coil, T2001, from collector to base. Oscillator tuning is done by T2001 and the capacitance selected by the function switch. For instance, when the VOLUME switch is pressed to the right, the entire capacitive circuit to the right of T2001 is across the coil, and the battery is connected. Notice that C2003 is adjustable so that the oscillator can be set to 44.750 kHz.

If the VOLUME switch is moved to the left, the same thing happens, except that now C2010 is switched in, changing the frequency to 38.750 kHz. The other two functions work in similar fashion.

The remote transmitter we have looked at was a four-function one. Some sets employ as many as eight remote functions. Fig. 17-15 shows an example of such a transmitter. You may find it interesting to go through this one yourself and see how the various frequencies for the different functions are determined.

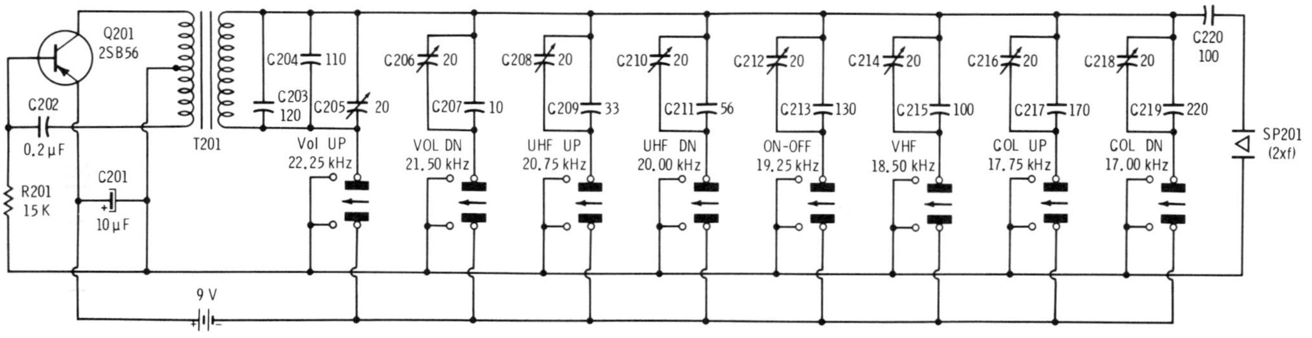

Fig. 17-15. An eight-function transmitter.

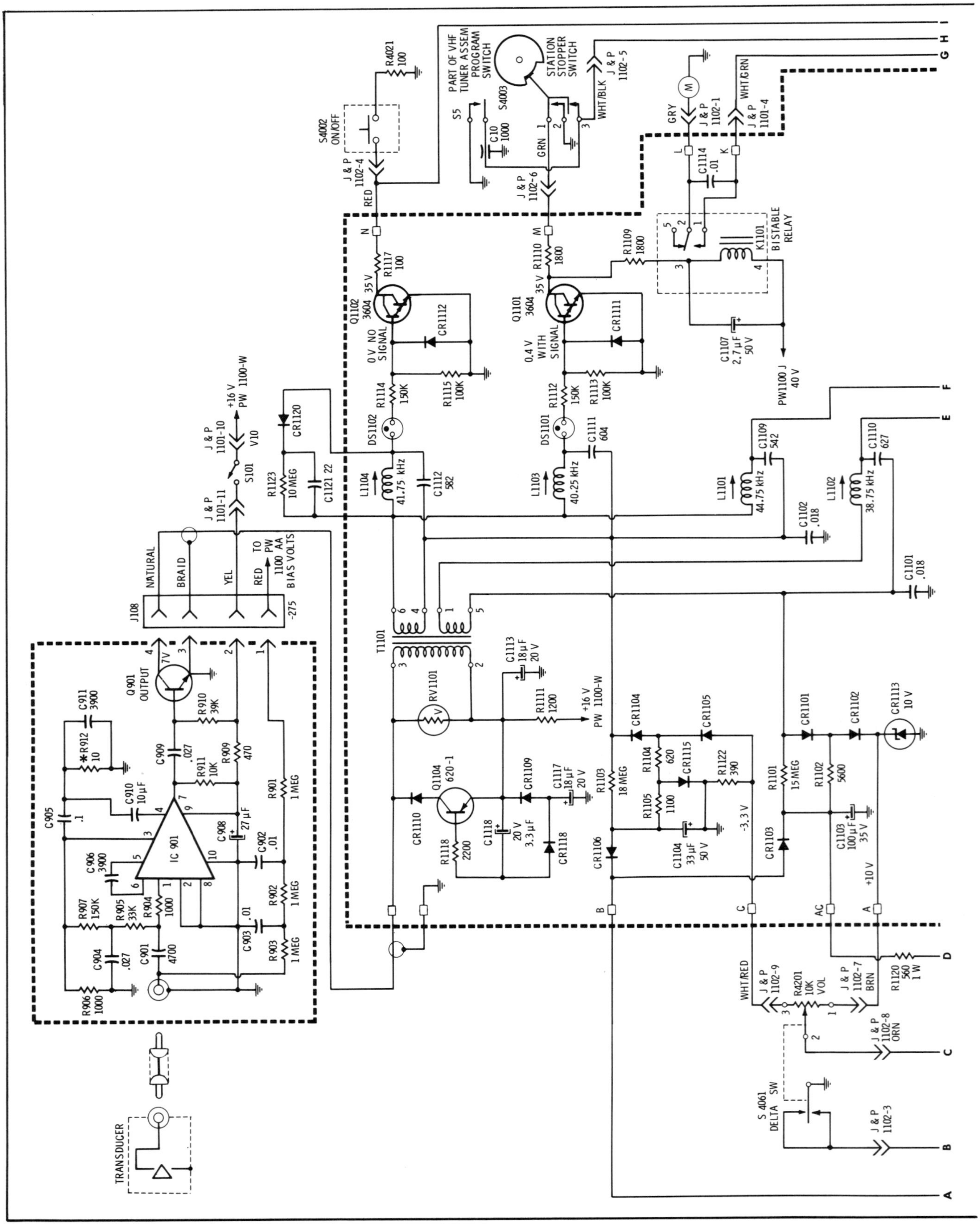

Fig. 17-16. Schematic diagram

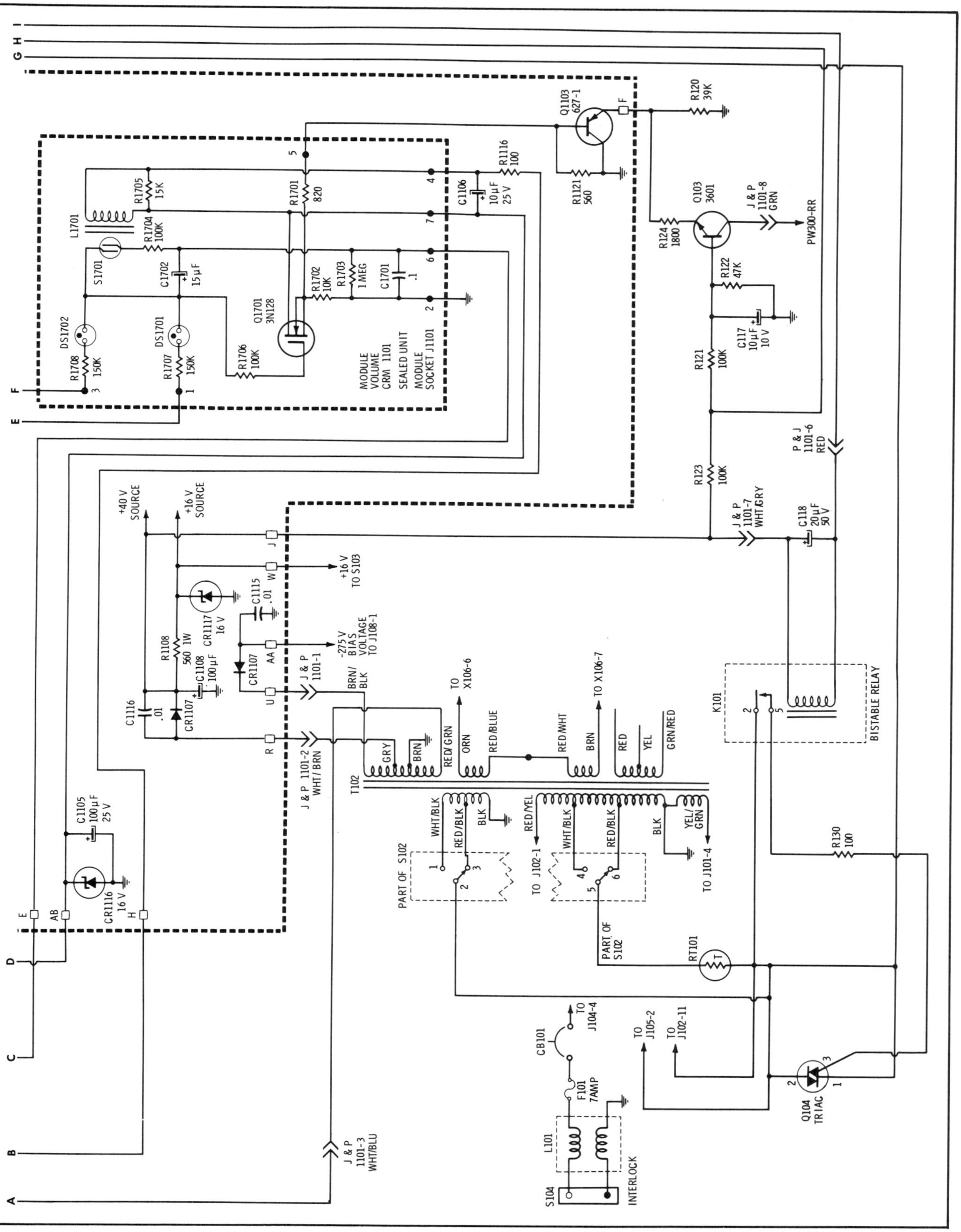

of a remote-control receiver.

17-10 Remote Control Receivers

The receiver in Fig. 17-16 is the one that is used with the four-function transmitter in Fig. 17-14. The incoming signal is picked up by the transducer and is amplified by IC901 and Q901. The signal is then adjusted to a constant level by Q1104 and the components associated with it. From the limiter, the signal is coupled into the filters through transformer T1101.

Suppose that the incoming signal is at 41.75 kHz. It will be coupled through filter L1104-C1112 to the on/off control (Q1102), which actuates the on/off relay (K101).

The remaining functions operate in much the same manner, with Q1101 controlling the channel-selector motor and CRM1101 controlling the volume.

17-11 Troubleshooting Remote Controls

Troubleshooting remote controls is not difficult, but it does require some understanding of how they work. Also, an oscilloscope is necessary for making tests. Let us use the four-function system shown in Figs. 17-14 and 17-16 as an example.

Suppose you have a set that works fine manually, but the remote control won't do anything. First, be sure the remote control is turned on. It would be frustrating to spend a lot of time troubleshooting a remote control that was switched off.

Next, check the transmitter. The first step here is to test the battery under load. If the battery is all right, try the transmitter and see what happens—try each function. If nothing happens, try it again with a transmitter of known accuracy, if you have one. If the remote control works with the substitute transmitter but not with the other one, you have a transmitter problem. If the remote controls don't work with *either* transmitter, you probably have a receiver problem.

This time, assume that you have a transmitter problem (or that you don't have a standard transmitter to use for testing). In such a case, you would check the output of an electronic transmitter by loosely coupling it into the oscilloscope and looking at the waveform. If the waveform is missing, you have a transmitter problem. If not, you have one of two problems; either the transmitter is out of tune or the receiver is dead.

In the case of a mechanical transmitter, all you can do is open it up and look for broken or damaged parts. The oscilloscope won't help—there is no electricity involved, so there is no electronic signal to couple into the scope.

The troubleshooting chart in Fig. 17-17 should help clear up the procedure for troubleshooting a remote transmitter.

If you determine that the trouble is in the receiver, you can proceed as follows. First connect the scope across the input to the filters. Then try the transmitter (or another known good one). If the signal is present, the transmitter must be out of tune (or *all* of the receiver function control circuits are dead—not very likely). If the signal isn't present, the trouble must be either a dead transmitter or trouble in the pickup, amplifier, or limiter.

17-12 One Remote Function Fails

As a final problem to consider, think about a situation in which everything works except one function. In this case, you have only two choices. Either the transmitter isn't working on that frequency (check it by connecting the scope across the output of the filter and trying the transmitter), or the control circuit for that function is dead (it can be checked in the usual ways).

17-13 Review Questions

1. What does acc mean? What does it do?
2. What is dacc and what does it do in a set?
3. The acc delay diode (E4S) in Fig. 17-4 is short-circuited. What symptoms would you expect on the screen? Why?
4. If diode E2S in Fig. 17-4 was open-circuited, how would the acc be affected?
5. What would be the symptoms of a shorted acc amplifier in Fig. 17-4?
6. What would be the cause of the symptoms shown in Fig. 17-5C?
7. The 8-μF capacitor in the emitter circuit of Q4L in Fig. 17-6 is shorted. What symptoms would you expect?
8. The brightness control in Fig. 17-6 varies the brightness on the screen but the abl range control does not. Which components would you suspect? Why?
9. A certain tv set operates all right as long as the aft defeat switch is closed, but when the switch is open, the picture is lost, leaving only severe snow. What would you do? (Use Fig. 17-8 as a basis for your answer.)
10. A set contains the fti circuit shown in Fig.

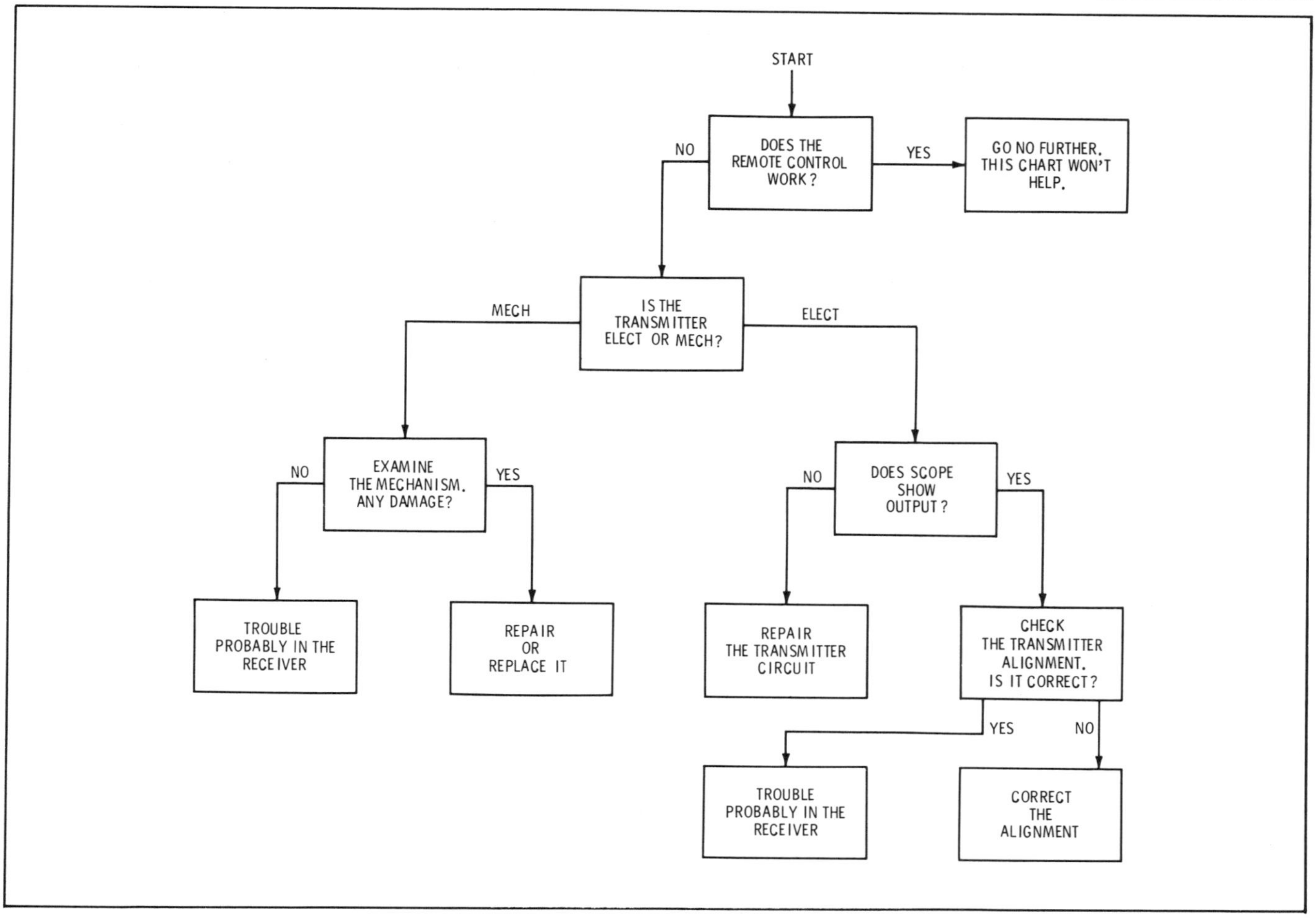

Fig. 17-17. Chart for troubleshooting the remote transmitter.

17-9. The fti lamp (E11) stays on all the time, but the picture is properly tuned. What would you do?

11. Transistor Q2 in Fig. 17-9 is found to be shorted from emitter to collector. Could this cause the symptoms in Question 10? Why?
12. The fti lamp in Fig. 17-9 will not light at all, even if the set is severely detuned. The lamp itself checks OK. What would you do next? Why?
13. What would the lamp in Fig. 17-9 do if the set was switched to an unused channel? Why?
14. The line regulator in Fig. 17-10 has a shorted triac. What would be the voltage across the power-transformer primary?
15. Why are line-voltage regulators used in some television sets?
16. Which of the sets at the end of the book have:

    A. Acc?
    B. Abl?
    C. Remote control?
    D. Aft?

17. Explain how the addition of automatic features to the modern tv set made remote control more convenient.
18. Are all remote controls in the 40-kHz range? What other frequencies did you find in this chapter?
19. What does "ultrasonic" mean?
20. Explain how a mechanical remote transmitter is tuned.
21. How does an electronic remote transmitter work?
22. Give two advantages of the mechanical type of transmitter.
23. Would you consider light weight an advantage of electronic remote transmitters? Explain your answer.
24. In an electronic remote transmitter, is there any battery drain while the unit is *not* in use? Why?
25. Why is it a good idea to have a known accu-

rate transmitter on hand? How is it used?

26. Explain briefly how a remote-control receiver works.
27. What would you check if *only* the off-on function of the receiver in Fig. 17-16 didn't work?
28. How do you align an electronic transmitter in a remote-control system?
29. How do you check to see if a mechanical transmitter is working?
30. How do you check to see if an electronic transmitter is working?
31. When you are following the steps in Fig. 17-17, what do you do if the transmitter is producing an output on all functions but the remote control doesn't work?
32. How do you proceed if only one function does not work?
33. What would you do if two out of four remote functions did not work?

CHAPTER 18

Troubleshooting Summary

In troubleshooting a television receiver, it is necessary to observe the symptoms carefully and classify them to organize your thinking into a logical pattern. The pattern used requires you to:

1. Be sure the set is connected and adjusted for normal viewing.
2. Observe and classify any abnormality as being sound, video, raster, or color trouble.
3. Localize the trouble to a single section of the set based on symptom classification.
4. Isolate the trouble to the smallest number of stages possible by observation.
5. Identify the defective component by comparing measurements to normal conditions.
6. Replace the defective component, and then repeat steps 1 and 2 to see if the set is operating properly.

By carrying out these steps, you can service any kind of television set failure. Examine each of these steps individually.

18-1 Connecting and Adjusting the Set

Before attempting to diagnose a problem, be very sure that the set is connected and adjusted for normal operation. To do this, you can follow the plan shown in Fig. 18-1. When you are sure that the set is ready, observe its performance.

18-2 Classifying Symptoms

With the set adjusted for normal viewing, listen to the sound and look at the picture. Observe and note any wrong condition in sound, video, raster, or color. Fig. 18-2 shows a systematic way of doing this. When the trouble has been identified as a sound, video, raster, or color problem, focus your attention on the section of the set that can cause it.

18-3 Localizing Troubles

When you have classified a problem as affecting *only* the sound, it can be localized right away to the sound section. This is illustrated in Fig. 18-3. Similarly, problems affecting *only* the color can be localized right away to the color section. Raster problems are harder to localize, as can be seen in Fig. 18-4. From this chart, you can see that problems with the raster may (depending on the type of symptoms) be localized to:

1. The picture tube

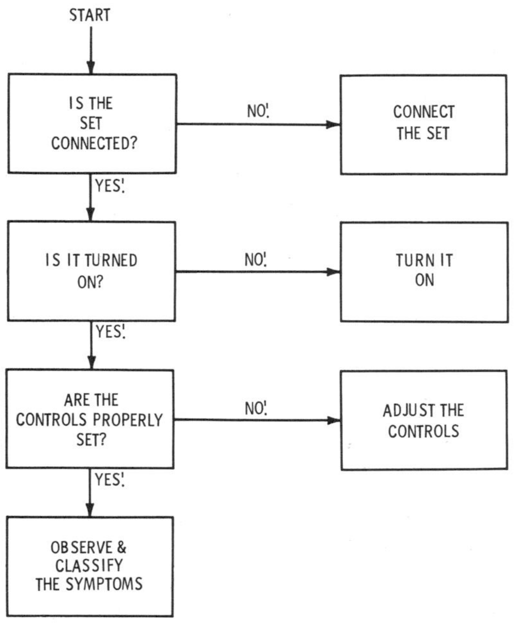

Fig. 18-1. Connecting and adjusting the set.

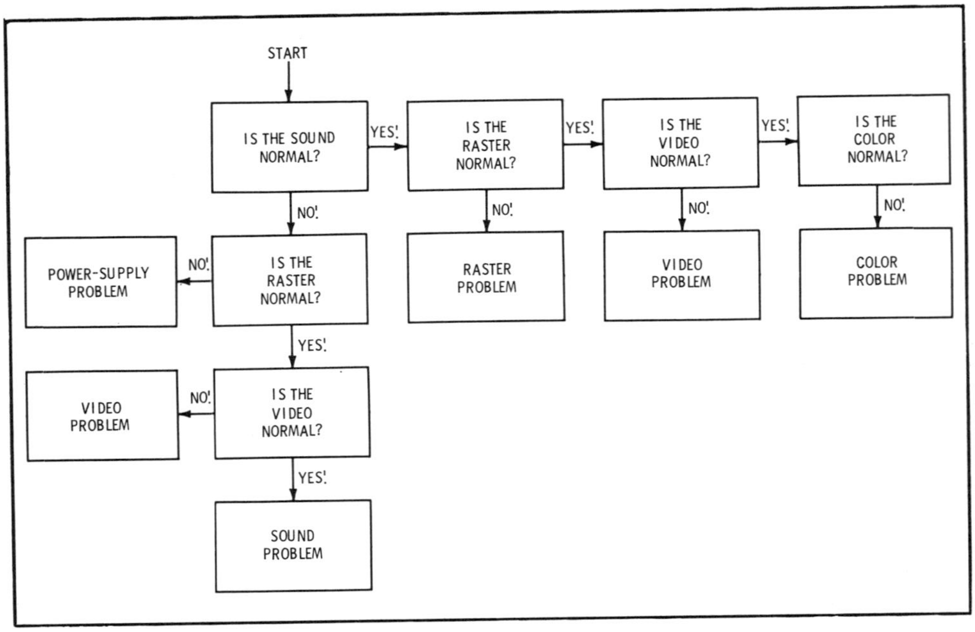

Fig. 18-2. Classifying symptoms.

2. The rf/if section
3. The power supplies
4. The sweep section

Similarly, you can see from Fig. 18-5 that localizing video problems can also be a little involved. You will notice that video problems can be localized to:

1. The picture tube
2. The rf/if section
3. The power supplies
4. The video section
5. The agc section

After the symptoms have been localized to a section of the set, further isolate them to the smallest number of stages possible.

18-4 Isolating Troubles Within a Section

Once a problem has been localized to a section within the set, use the controls and observations

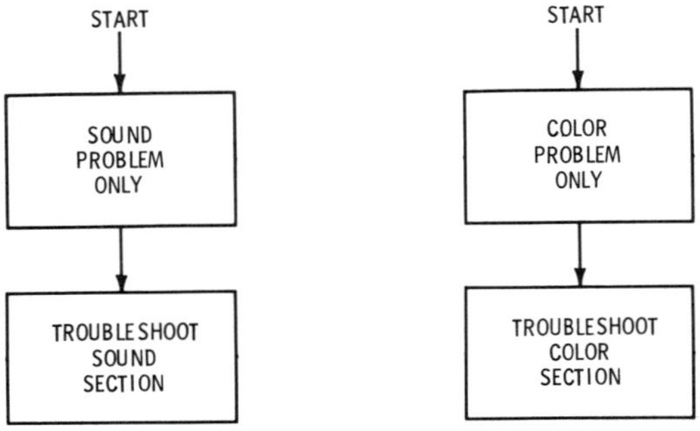

Fig. 18-3. Localizing sound and color problems.

to isolate it as much as possible. So far, eight sections of the set have been considered. They were:

1. The sound section
2. The picture tube
3. The video section
4. The rf/if section
5. The agc section
6. The sweep section
7. The power supplies
8. The color section

Let's take each one in turn and consider how troubles can be further isolated.

In the sound section, the volume control is normally located in the first audio-amplifier stage. Fig. 18-6 shows the block diagram of a typical set. The sound if and detector stages come before the volume control. The audio amplifiers, on the other hand, come after the volume control. This fact can be used to help isolate sound problems. Fig. 18-7 shows how this can be done.

The picture tube is normally checked with a picture-tube tester if it is suspected of having failed. In a color set, picture-tube trouble normally appears as a color trouble, rather than as a video problem.

The video section contains the contrast control and two or three video-amplifier stages. Fig. 18-8 shows how the video section can be checked.

The rf/if section of the set can be dealt with by carefully looking for snow in the picture and listening for a "hissing" noise in the sound. Fig.

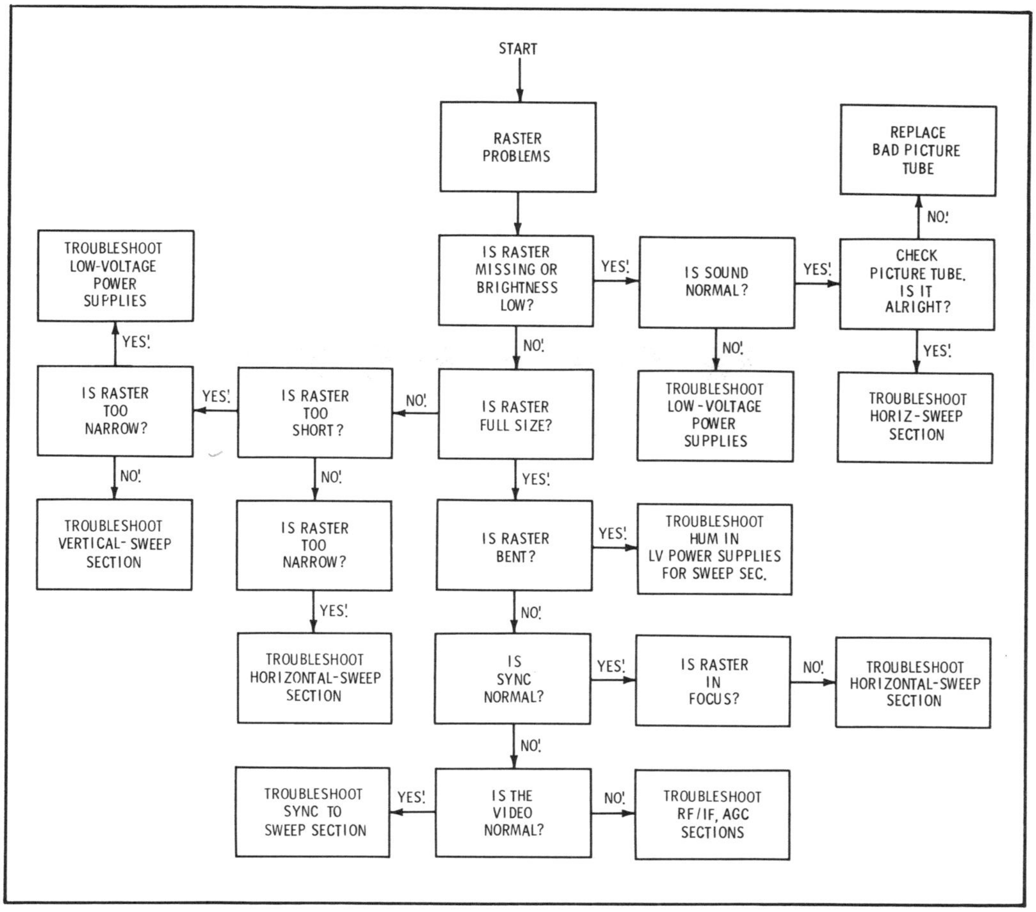

Fig. 18-4. Localizing raster problems.

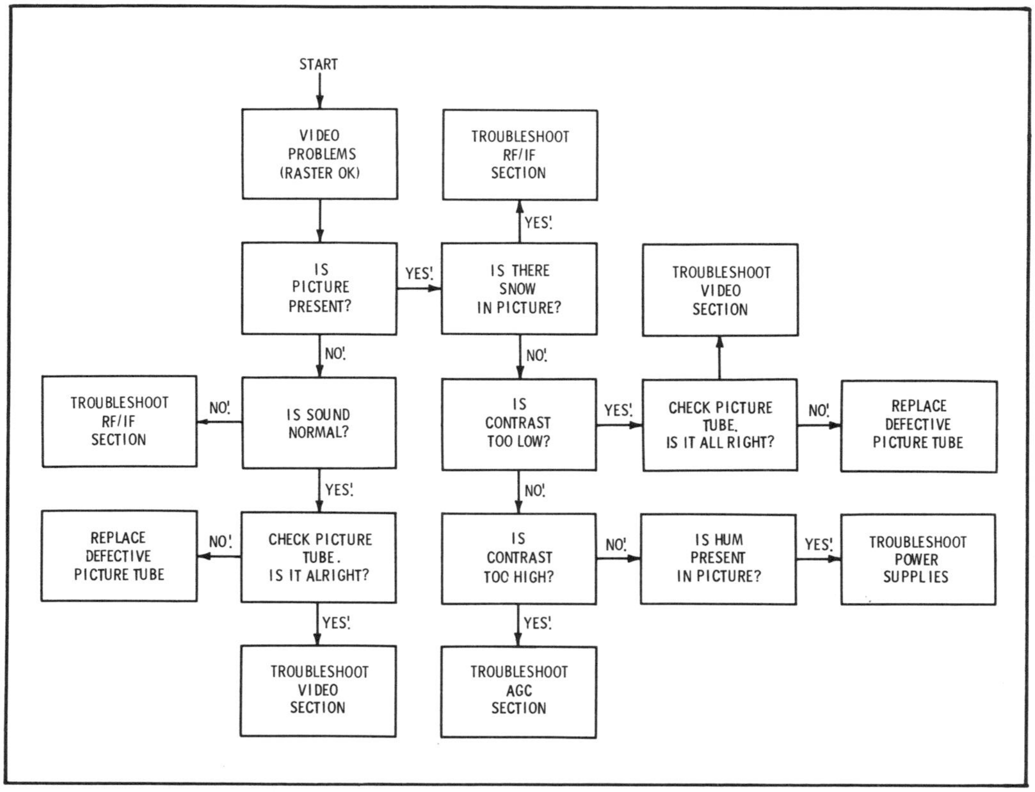

Fig. 18-5. Localizing video problems.

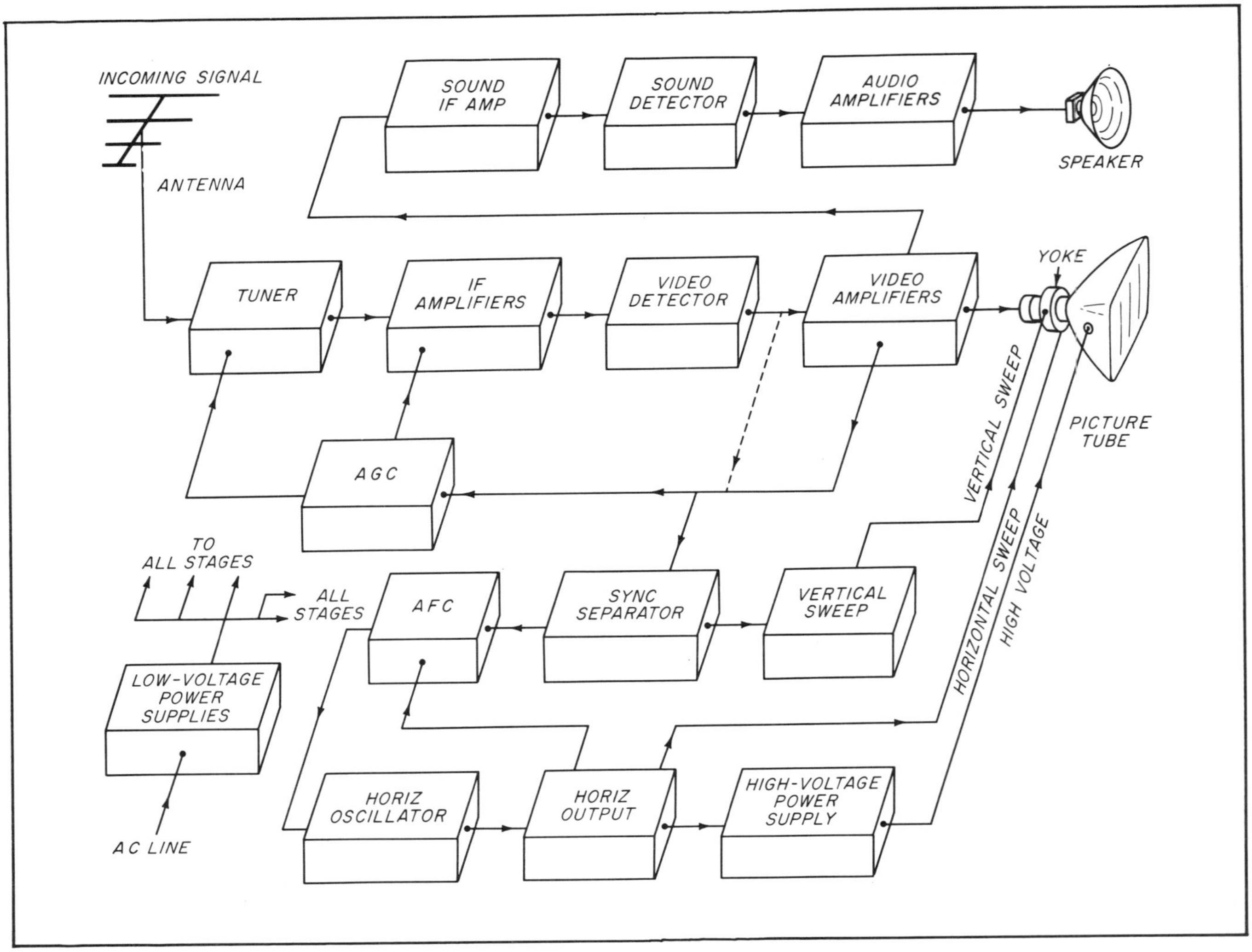

Fig. 18-6. Block diagram of typical black-and-white receiver.

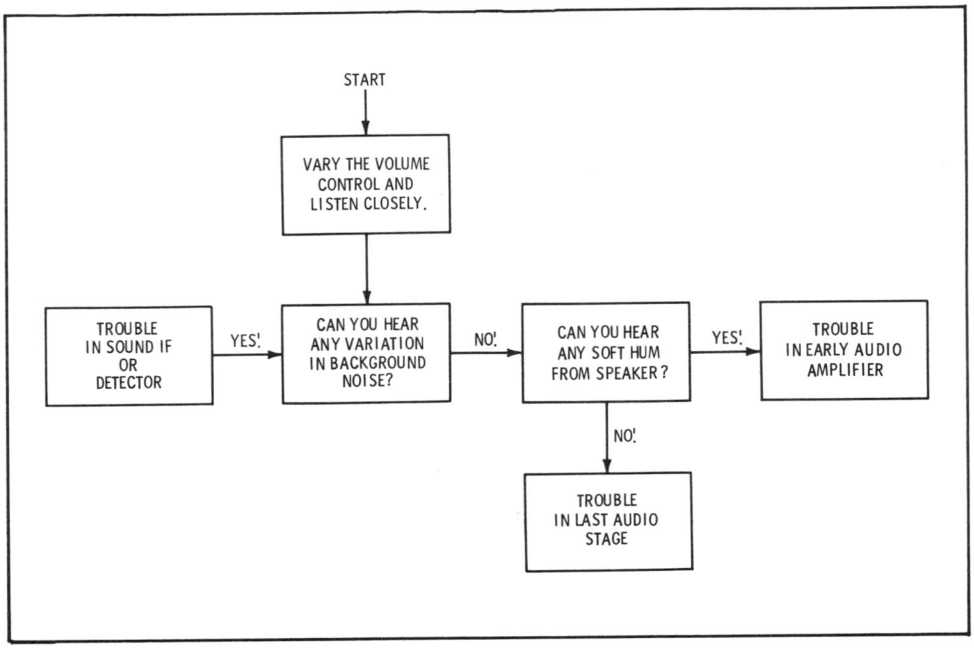

Fig. 18-7. Isolating sound problems.

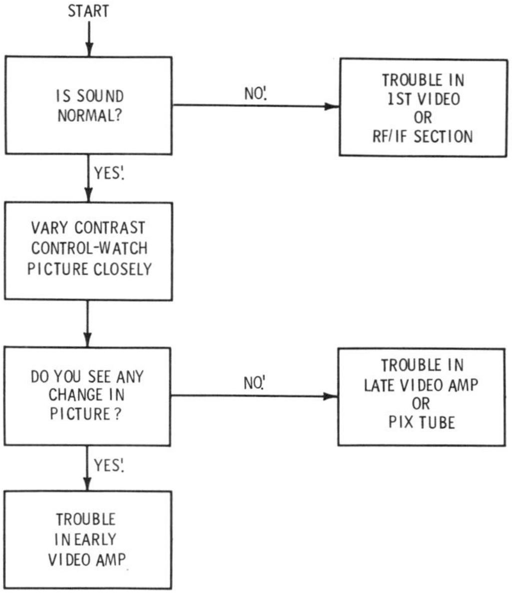

Fig. 18-8. Isolating video-section trouble.

18-9 shows the process that can be used to localize rf/if problems.

Trouble in the agc section can cause the picture to have either too much contrast (and be bent) or not enough contrast (and have weak sync). There is normally an agc control that can be used to help isolate agc problems. Fig. 18-10 shows the procedure.

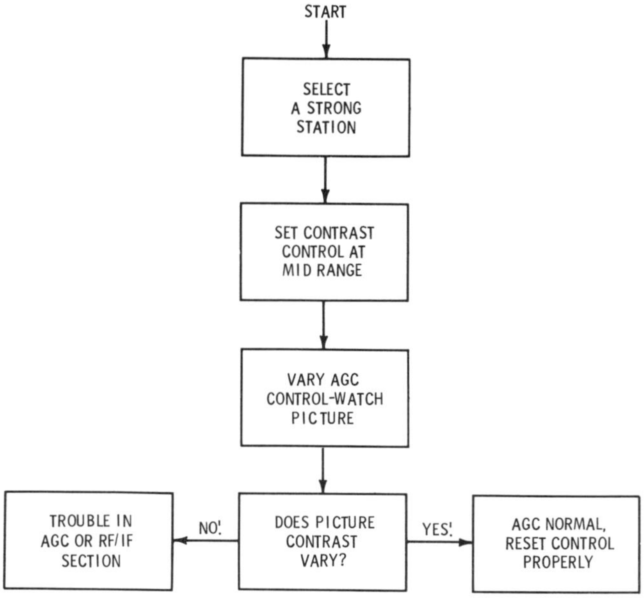

Fig. 18-10. Isolating agc troubles.

The sweep section of a television set has four basic functions. These basic functions are:

1. Synchronizing the picture
2. Sweeping the picture vertically
3. Sweeping the picture horizontally
4. Producing the picture-tube high voltage

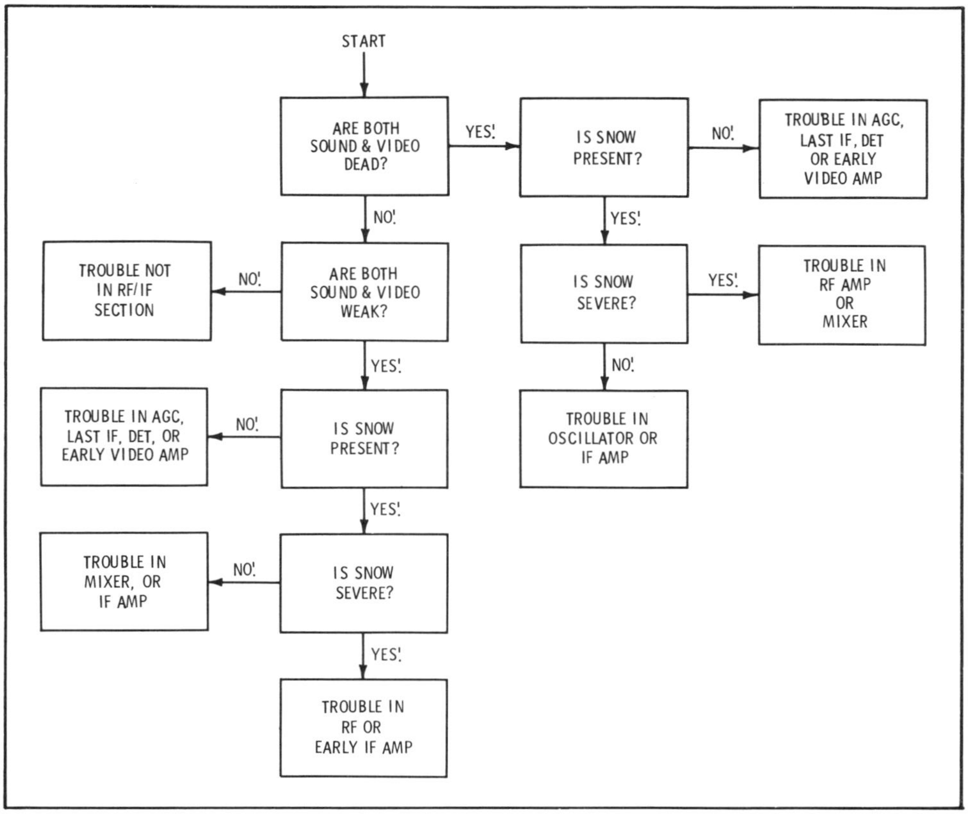

Fig. 18-9. Isolating rf/if troubles.

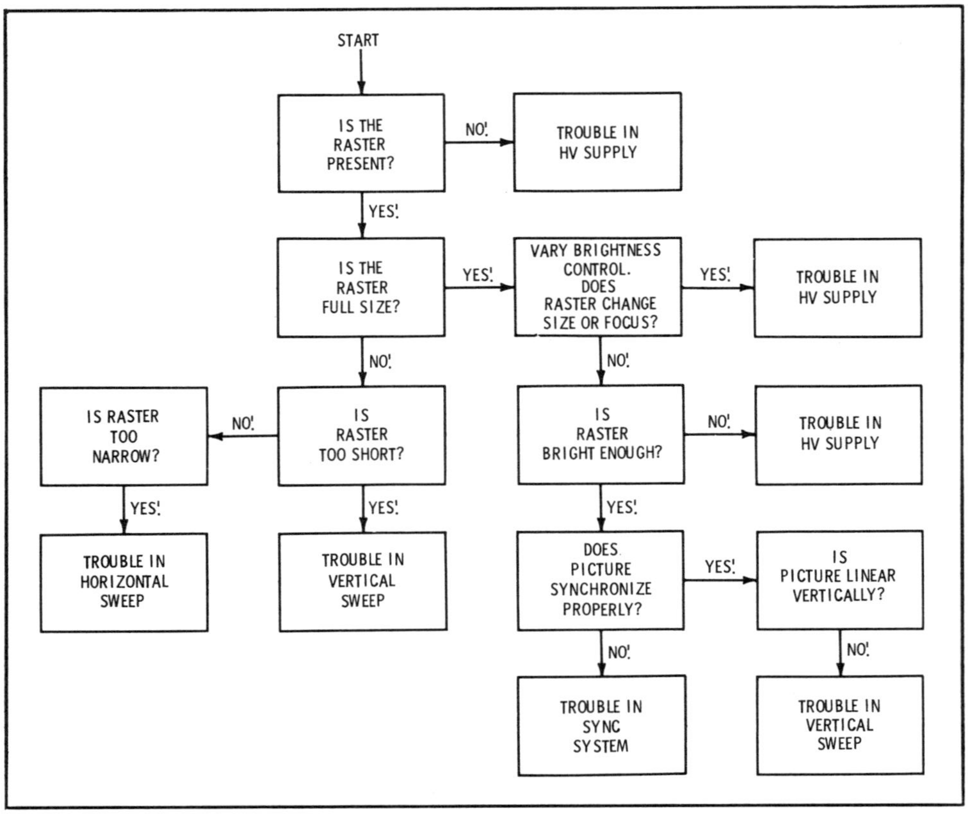

Fig. 18-11. Identifying sweep-function failures.

In isolating sweep problems, you must first identify which of the functions is not working properly. Fig. 18-11 gives a chart for doing this. When you know which sweep-section function has failed, you can proceed to localize the trouble within the function.

Figs. 18-12, 18-13, 18-14, and 18-15 show processes that can be used to localize most sweep-section problems.

The low-voltage power supplies can cause troubles in any stage or in any combination of stages. The classic symptoms of low-voltage power-supply failures are shown in Fig. 18-16.

In isolating color problems, it is well to refer to

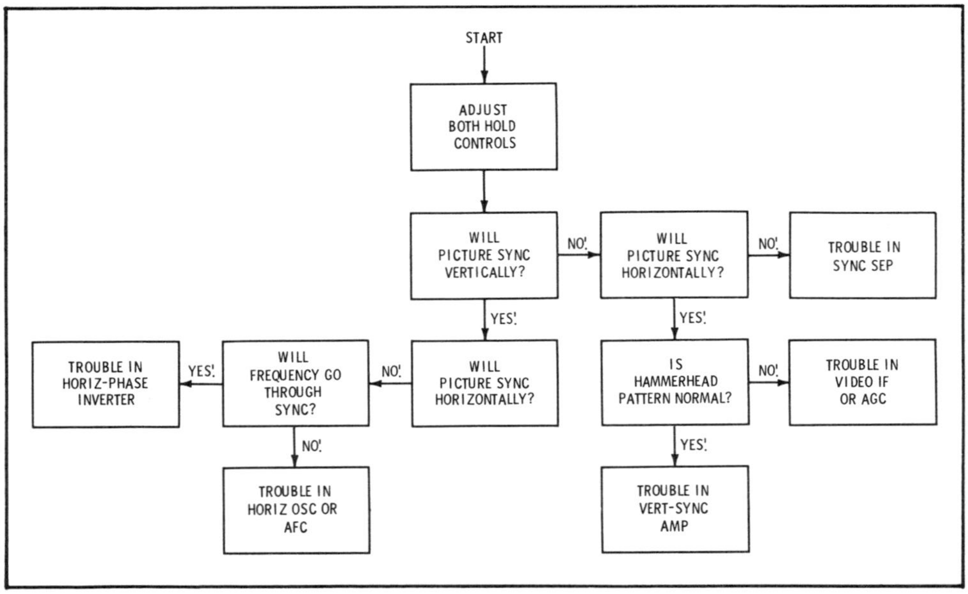

Fig. 18-12. Isolating sync problems.

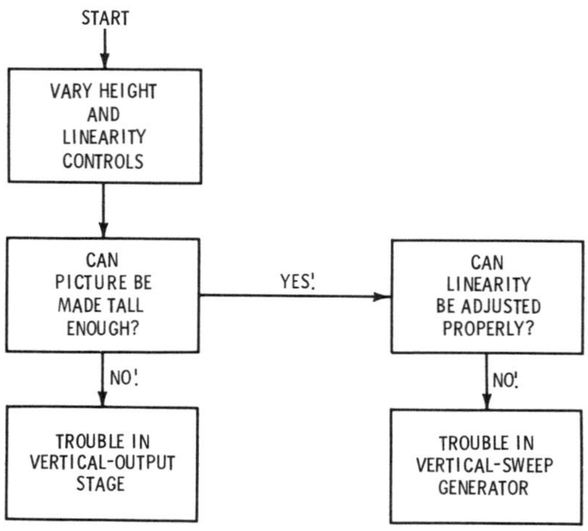

Fig. 18-13. Isolating vertical-sweep problems.

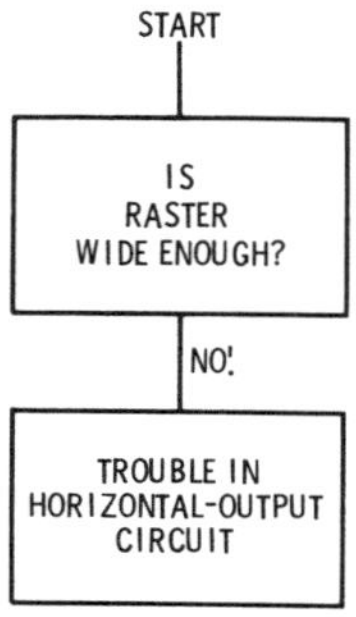

Fig. 18-14. Isolating horizontal-sweep problems.

one of the block diagrams of a color set. There are four basic types of color problems. These types of problems are:

1. Loss of all color
2. Loss of one color
3. One excessive color
4. Loss of color sync

An approach to isolating these four types of problems is given in Fig. 18-17.

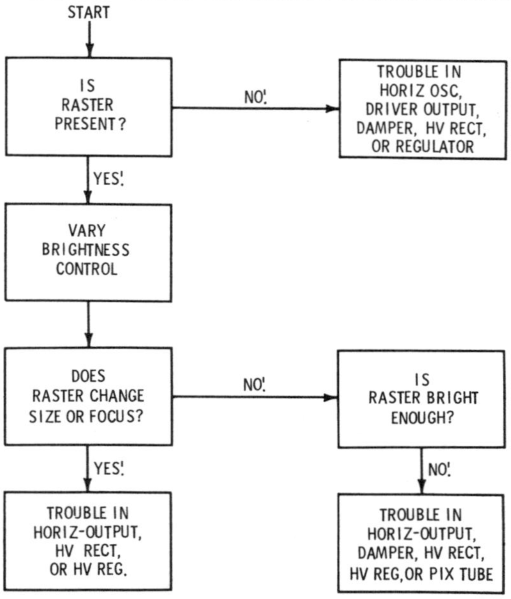

Fig. 18-15. Isolating high-voltage supply problems.

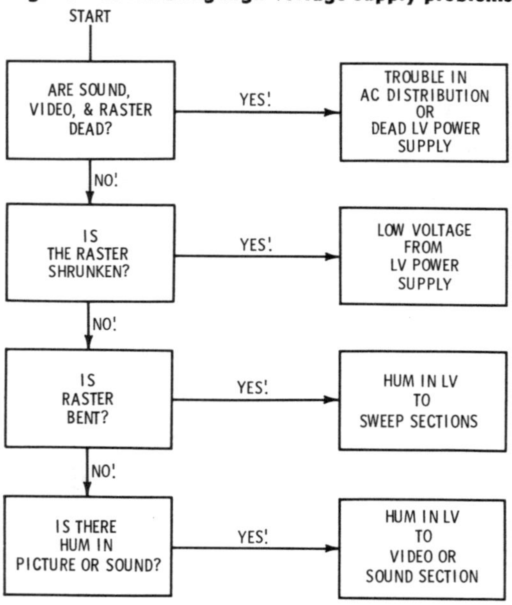

Fig. 18-16. Isolating low-voltage supply failures.

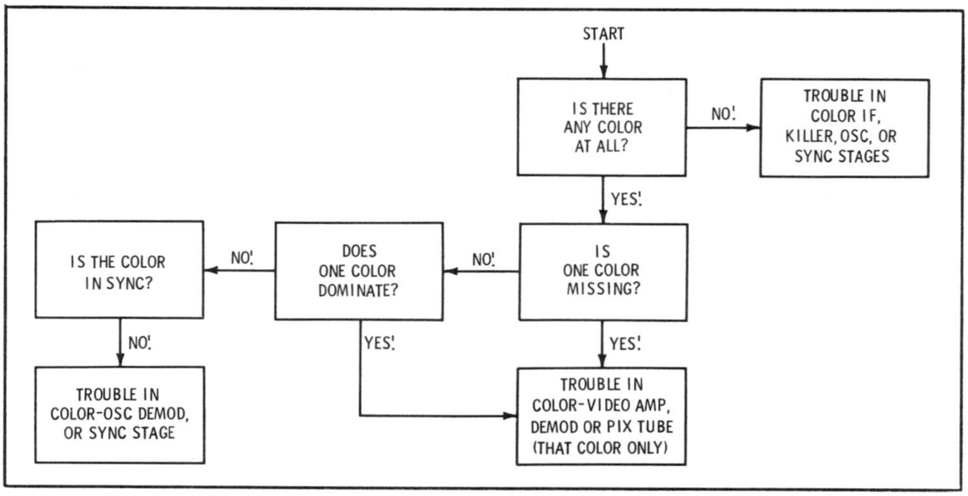

Fig. 18-17. Isolating color problems.

18-5 Identifying Defective Components

When a problem has been isolated as much as possible, go into the circuitry to locate the component at fault. This can be most effectively done by measuring circuit voltages and observing circuit waveforms. The results of these measurements and observations can then be compared with those given on a manufacturer's circuit diagram or in a Sams PHOTOFACT® folder.

18-6 Practice Exercises

On the following pages are several sets of symptoms. Use them to practice diagnosing television problems.

Fig. 18-18. Exercise 1.

Symptoms			
Sound	**Raster**	**Video**	**Color**
Normal.			
Observations:			
Diagnosis:			

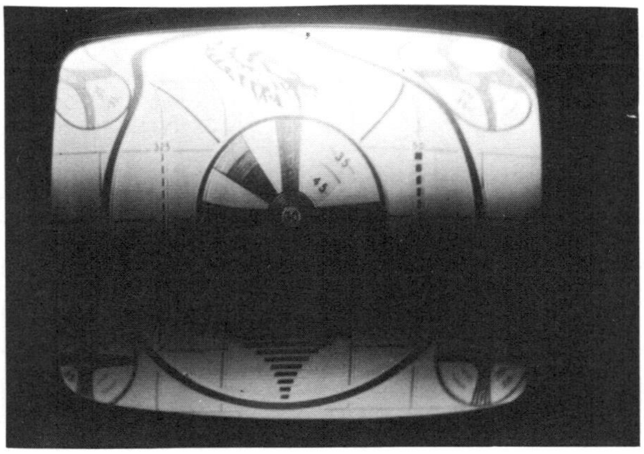

Fig. 18-19. Exercise 2.

Sound	Raster	Video	Color
Hum in sound.			
Observations			
Diagnosis:			

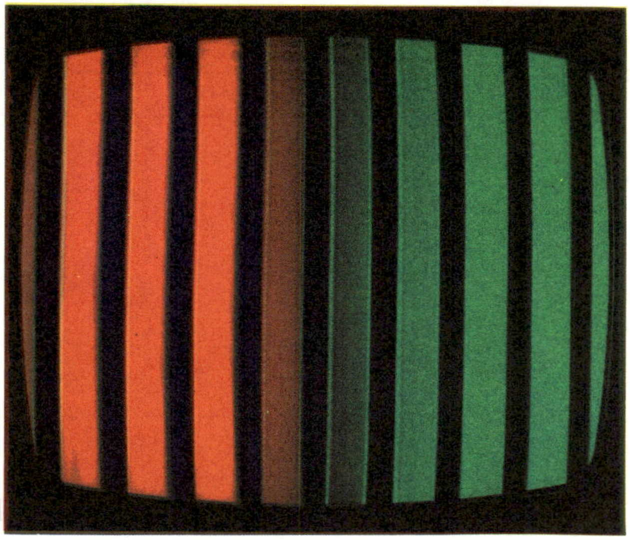

Fig. 18-20. Exercise 3.

Symptoms			
Sound	**Raster**	**Video**	**Color**
Normal.			

Observations: Screen controls will not correct symptom. Otherwise normal.

Diagnosis:

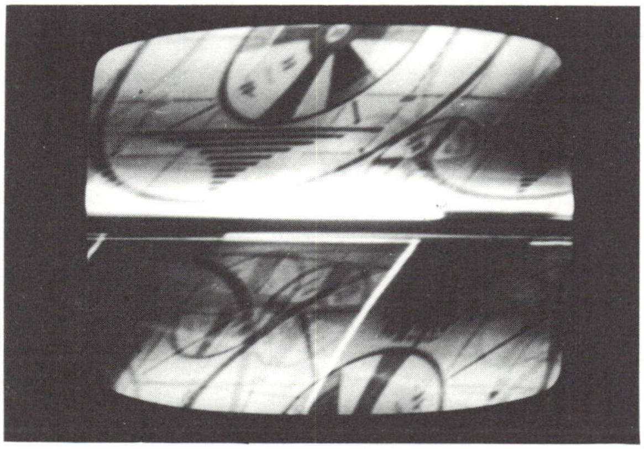

Fig. 18-21. Exercise 4.

Sound	Raster	Video	Color
Normal.			

Observations: Hold controls correct but will not lock in.

Diagnosis:

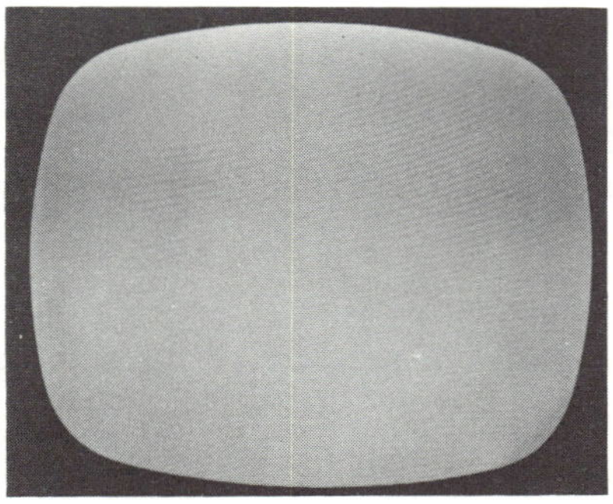

Fig. 18-22. Exercise 5.

Sound	Raster	Video	Color
Dead.			

Observations: Some background noise. It changes with volume setting.

Diagnosis:

Fig. 18-23. Exercise 6.

Symptoms

Sound	Raster	Video	Color
Normal.			

Observations:

Diagnosis:

Fig. 18-24. Exercise 7.

Sound	Raster	Video	Color
Normal.			

Observations:

Diagnosis:

Fig. 18-25. Exercise 8.

Sound	Raster	Video	Color
Noise only.			

Observations: Tuning is set correctly.

Diagnosis:

Fig. 18-26. Exercise 9.

Symptoms			
Sound	**Raster**	**Video**	**Color**
Dead.			

Observations: Set is plugged in and turned on.

Diagnosis:

Fig. 18-27. Exercise 10.

Sound	Raster	Video	Color
Normal.			

Observations: Screen controls setting normal.

Diagnosis:

Fig. 18-28. Exercise 11.

Sound	Raster	Video	Color
Normal.			

Observations: Hold control will not correct.

Diagnosis:

Fig. 18-29. Exercise 12.

Symptoms

Sound	Raster	Video	Color
Normal.			

Observations: Controls set correctly.

Diagnosis:

Fig. 18-30. Exercise 13.

Sound	Raster	Video	Color
Normal.			

Observations: All controls are properly set.

Diagnosis:

Fig. 18-31. Exercise 14.

Sound	Raster	Video	Color
Dead.			

Observations: Slight change in background as volume is varied.

Diagnosis:

Fig. 18-32. Exercise 15.

Symptoms

Sound	Raster	Video	Color
Normal.			

Observations: Hor. hold varies frequency through sync but will not hold.

Diagnosis:

Fig. 18-33. Exercise 16.

Sound	Raster	Video	Color
Normal.			

Observations: Hold control has very little effect.

Diagnosis:

Fig. 18-34. Exercise 17.

Sound	Raster	Video	Color
Normal.			

Observations: Agc will not correct.

Diagnosis:

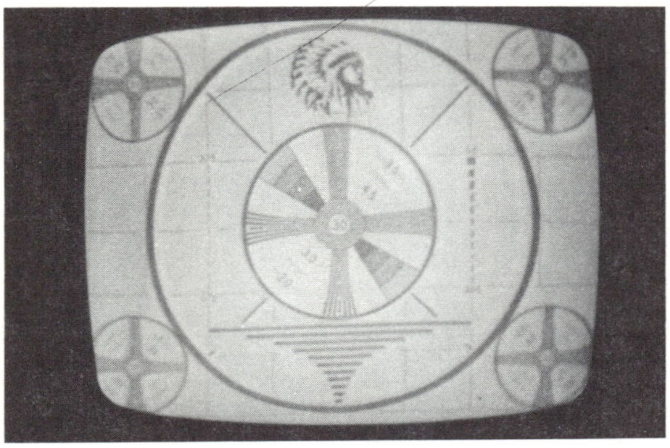

Fig. 18-35. Exercise 18.

Symptoms			
Sound	**Raster**	**Video**	**Color**
Normal.			
Observations: Contrast control at maximum. Pix tube checks OK.			
Diagnosis:			

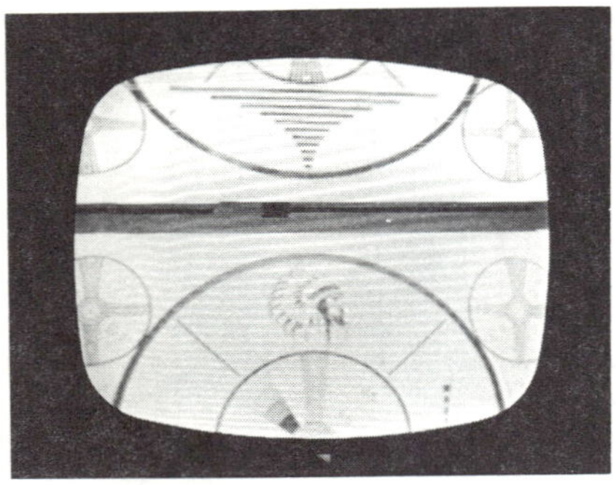

Fig. 18-36. Exercise 19.

Sound	Raster	Video	Color
Normal.			
Observations: Hold control will not lock in picture.			
Diagnosis:			

Fig. 18-37. Exercise 20.

Sound	Raster	Video	Color
Dead.			
Observations: No sound at all from speaker when volume is varied.			
Diagnosis:			

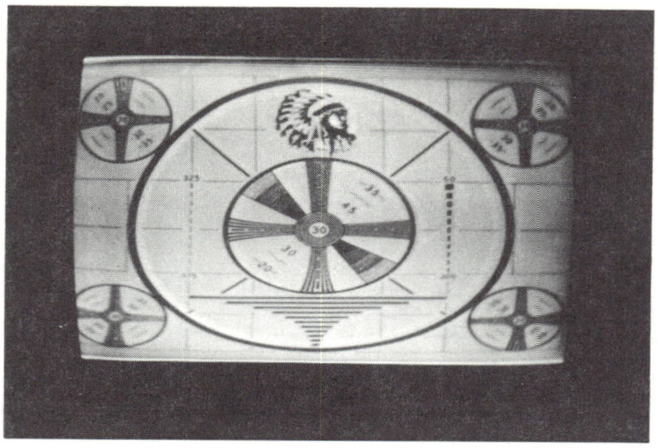

Fig. 18-38. Exercise 21.

Symptoms			
Sound	**Raster**	**Video**	**Color**
Normal.			
Observations:			
Diagnosis:			

Fig. 18-39. Exercise 22.

Sound	Raster	Video	Color
Normal.			
Observations:			
Diagnosis:			

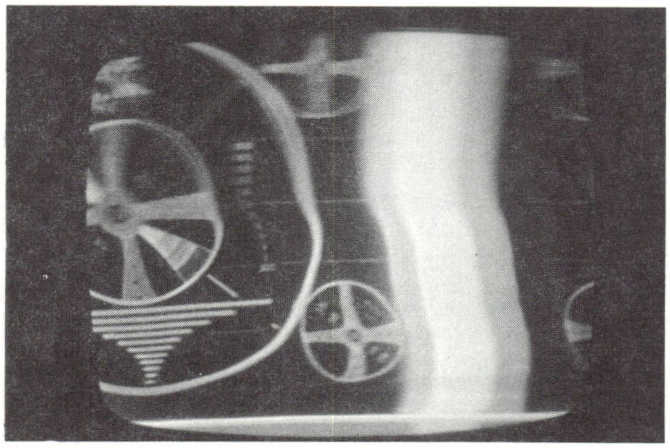

Fig. 18-40. Exercise 23.

Sound	Raster	Video	Color
Severe buzz in sound.			
Observations:			
Diagnosis:			

Fig. 18-41. Exercise 24.

Symptoms			
Sound	**Raster**	**Video**	**Color**
Normal.			

Observations: Screen controls settings are OK.

Diagnosis:

FOLDOUT SECTION

Circuit Diagrams